AGRIBUSINESS

Marketing and Management

AGRIBUSINESS

Marketing and Management

Honey John

Aarthi Aishwarya

Priyadharshini M

KNOWLEDGE
BAKERS

KNOWLEDGE BAKERS

ISBN: 978-93-49964-19-8

Published by : **Knowledge Bakers**
Pune, India
Email: knowledgebakers2019@gmail.com

PREFACE

Agribusiness: Marketing and Management is a comprehensive guide that delves into the strategic and operational aspects of marketing and managing agribusinesses in a rapidly evolving global landscape. As agriculture shifts from traditional practices to dynamic, market-oriented systems, the need for effective marketing strategies and sound management practices has become more critical than ever.

This book is intended for students, educators, agribusiness professionals, and policymakers seeking to understand the core principles that drive success in the agricultural value chain—from farm to fork. It integrates key concepts of business management, marketing, supply chain logistics, and consumer behavior within the specific context of agricultural enterprises. Whether it's managing farm operations, branding agri-products, or navigating market access and competition, this book offers practical frameworks and real-world insights.

Chapters address topics such as agribusiness planning, value addition and product differentiation, distribution channels, pricing strategies, market research, organizational behavior, and decision-making under uncertainty. Case studies and applied examples are used throughout to bridge theory with practice, making the material relevant and engaging.

We extend our sincere appreciation to the many contributors—researchers, farmers, industry leaders, and institutions—who have informed the development of this volume. It is our hope that this book not only serves as a foundational resource but also encourages innovation, sustainability, and resilience in the marketing and management of agribusinesses around the world.

Honey John

Aarthi Aishwarya

Priyadharshini M

CONTENTS

1

AGRIBUSINESS: OVERVIEW

Agribusiness is the business sector encompassing farming and farming-related commercial activities. It involves all the steps required to send an agricultural good to market, namely production, processing, and distribution. This industry is an important component of the economy in countries with arable land since agricultural products can be exported.

Agribusiness treats the different aspects of raising agricultural products as an integrated system. Farmers raise animals and harvest fruits and vegetables with the help of sophisticated harvesting techniques, including the use of GPS to direct operations. Manufacturers develop increasingly efficient machines that can drive themselves. Processing plants determine the best way to clean and package livestock for shipping. While each subset of the industry is unlikely to interact directly with the consumer, each is focused on operating efficiently in order to keep prices reasonable.

Agribusiness refers to the enterprises, the industry, and the field of study of value chains in agriculture and in the bio-economy, in which case it is also called bio-business or bio-enterprise. The primary goal of agribusiness is to maximize profit while sustainably satisfying the needs of consumers for products related to natural resources such as biotechnology, farms, food, forestry, fisheries, fuel, and fiber — usually with the exclusion of non-renewable resources such as mining.

Studies of business growth and performance in farming have found successful agricultural businesses are cost-efficient internally and operate in favorable economic, political, and physical-organic environments. They are able to expand and make profits, improve the productivity of land, labor, and capital, and keep their costs down to ensure market price competitiveness.

Agribusiness is not limited to farming. It encompasses a broader spectrum through the agribusiness system which includes input supplies, value-addition, marketing, entrepreneurship, microfinancing, agricultural extension, among others.

In some countries like the Philippines, creation and management of agribusiness enterprises require consultation with registered agriculturists if reached a certain level of operations, capitalization, land area, or number of animals in the farm.

UNDERSTANDING AGRIBUSINESS

Market forces have a significant impact on the agribusiness sector, as do natural forces, such as changes in the earth's climate. Changes in consumer taste alter what products are grown and raised. For example, a shift in consumer tastes away from red meat may cause demand—and therefore prices—for beef to fall, while increased demand for produce may shift the mix of fruits and vegetables that farmers raise. Businesses unable to rapidly change in accordance with domestic demand may look to export their products abroad. If that fails, they may not be able to compete and remain in business.

Climate change has placed intensifying pressure on many companies in the agribusiness industry to remain relevant, and profitable, while adapting to the threats posed by large-scale shifts in weather patterns.

Agribusiness Challenges

Countries with farming industries face consistent pressures from global competition. Products such as wheat, corn, and soybeans tend to be similar in different locations, making them commodities. Remaining competitive requires agribusinesses to operate more efficiently, which can require investments in new technologies, new ways of fertilizing and watering crops, and new ways of connecting to the global market.

Global prices of agricultural products may change rapidly, making production planning a complicated activity. Farmers may also face a reduction in usable land as suburban and urban areas expand into their regions.

Use of New Technology

The use of new technology is vital to remain competitive in the global agribusiness sector. Farmers need to reduce crop costs and increase yield per square acre to remain competitive.

New drone technology is at the cutting edge of the industry. An article published in 2016 by the Massachusetts Institute of Technology (MIT) identified Six Ways

Drones Are Revolutionizing Agriculture. These techniques, including soil and field analysis, planting, and crop monitoring, will be key to improving crop yields and moving the agribusiness sector forward.

Key areas of concern for the use of drone technology remain the safety of drone operations, privacy issues, and insurance-coverage questions.

Agribusiness Examples

Agribusiness is a broad industry, it incorporates a wide range of different companies and operations. Agribusinesses include small family farms and food producers up to multinational conglomerates involved in the production of food on a national scale.

Some examples of agribusinesses include farm machinery producers such as Deere & Company, seed and agrichemical manufacturers such as Monsanto, food processing companies such as Archer Daniels Midland Company, as well as farmer's cooperatives, agritourism companies, and makers of biofuels, animal feeds, and other related products.

Evolution of the agribusiness concept

The word "agribusiness" is a portmanteau of the words agriculture and business. The earliest known use of the word was in the Volume 155 of the Canadian Almanac & Directory published in 1847. Although most practitioners recognize that it was coined in 1957 by two Harvard Business School professors, John Davis and Ray Goldberg after they published the book "A Concept of Agribusiness."

"Agribusiness is the sum total of all operations involved in the manufacture and distribution of farm supplies; production operations on the farm; and the storage, processing, and distribution of farm commodities and items made from them." (Davis and Goldberg, 1956) Their book argued against the New Deal programs of then U.S. President Franklin Roosevelt as it led to the increase in agricultural prices. Davis and Goldberg favored corporate-driven agriculture or large-scale farming to revolutionize the agriculture sector, lessening the dependency on state power and politics. They explained in the book that vertically integrated firms within the agricultural value chains have the ability to control prices and where they are distributed. Goldberg then assisted in the establishment of the first undergraduate program in agribusiness in 1966 at the UP College of Agriculture in Los Baños, Philippines as Bachelor of Science in Agriculture major in Agribusiness.

The program was initially a joint undertaking with the UP College of Business Administration in Diliman, Quezon City until 1975. Dr. Jose D. Drilon of the

University of the Philippines then published the book "Agribusiness Management Resource Materials" (1971) which would be the foundation of current agribusiness programs around the world. In 1973, Drilon and Goldberg further expanded the concept of agribusiness to include support organizations such as governments, research institutions, schools, financial institutions, and cooperatives within the integrated Agribusiness System.

Mark R. Edwards and Clifford J. Shultz II (2005) of Loyola University Chicago reframed the definition of agribusiness to emphasize its lack of focus on farm production but towards market centricity and innovative approach to serve consumers worldwide.

"Agribusiness is a dynamic and systemic endeavor that serves consumers globally and locally through innovation and management of multiple value chains that deliver valued goods and services derived from sustainable orchestration of food, fiber and natural resources." (Edwards and Shultz, 2005)

In 2012, Thomas L. Sporleder and Michael A. Boland defined the unique economic characteristics of agribusiness supply chains from industrial manufacturing and service supply chains. They have identified seven main characteristics:

Risks emanating from the biological nature of agrifood supply chains

The role of buffer stocks within the supply chain

The scientific foundation of innovation in production agriculture having shifted from chemistry to biology

Cyberspace and information technology influences on agrifood supply chains

The prevalent market structure at the farm gate remains oligopsony

Relative market power shifts in agrifood supply chains away from food manufacturers downstream to food retailers

Globalization of agriculture and agrifood supply chains

In 2017, noting the rise of genetic engineering and biotechnology in agriculture, Goldberg further expanded the definition of agribusiness which covers all the interdependent aspects of the food system including medicine, nutrition, and health. He also emphasized the responsibility of agribusiness to be environmentally and socially conscious towards sustainability.

"Agribusiness is the interrelated and interdependent industries in agriculture that supply, process, distribute, and support the products of agriculture." (Goldberg, 2017)

Some agribusinesses have adopted the triple bottom line framework such as aligning for fair trade, organic, good agricultural practices, and B-corporation certifications towards the concept of social entrepreneurship.

AGRIBUSINESS: VAST AND COMPLEX SYSTEM

Agriculture has evolved into agribusiness and has become a vast and complex system that reaches for beyond the farm to include all those who are involved in bringing food and fiber to consumers. Agribusiness include not only those that farm the land but also the people and firms that provide the inputs like Seed, chemicals, credit etc., process the output like Milk, grain, meat etc., manufacture the food products like icc-cream, bread, breakfast cereals etc., and transport and sell the food products to consumers as restaurants, supermarkets.

Initially agriculture being the major venture it was easy to become a farmer, but yield was low. Average Indian farmer produced enough food to feed just four people. As a consequence most farmers were nearly totally self-sufficient. They produced most of the inputs they needed for production, such as seed, draft animals, feed and simple farm equipment. Farm families processed the commodities they grew to make their own food and clothing. They consumed or used just about everything they produced.

The small amount of output not consumed on the farm was sold for cash. These items were used to feed and cloth the minor portion of the country's population that lived in villages and cities. A few agricultural products made their way into the export market and were sold to buyers is other countries.

Farmers found it increasingly profitable to concentrate on production and began to purchase inputs they formerly made themselves. This trend enabled others to build business that focused on meeting the need for inputs used in production agriculture such as seed, fencing, machinery and so on.

These farms involved into the industries that make up the "agricultural inputs sector".

At the same time, the agriculture input sector was evolving, a similar evaluation was taking place a commodity processing and food manufacturing moved off the farm. The form of most commodities like wheat, rice, milk, livestock and so on that must be changed to make them more useful and convenient for consumers. For ex. consumers would rather buy flour than grind the wheat themselves before backing a cake. They are willing to pay extra for the convenience of buying the processed commodity flour instead of the raw agriculture commodity like wheat.

During the same period technological advance were being made in food preservation method. Up until this time the perishable nature of most agriculture commodities meant that they were available only at harvest. Advance in food processing have made it possible to get those commodities all throughout the year. Today, even most farm families use purchased food and fiber products rather than doing the processing themselves. The farms that meet the consumers demand for greater processing and convenience also constitute a major part of agribusiness and are referred to as the processing manufacturing sector.

The evolution from agriculture to agribusiness has brought with it numerous benefits. These include reduced drudgery for laborers; the release of workers for nonagricultural endeavors; a better quality of food and fibers; a greater variety of products; improved nutrition; and increased mobility of people. The release of farm manpower and the creation of new, off-the-farm jobs have been the basis for the country's economic growth and development for the last 150 years. The key to this growth and development has been increased worker productivity, which in turn spurs creativity, new products and wealth. This translates into risk capital, new factories, new jobs, and increased consumer purchasing power.

It is apparent that the definition of agriculture had to be expanded to include more than production. Farmers rely on the input industries to provide the products and service they need to produce agricultural commodities. They also rely on commodity processors, food manufactures, and ultimately food distributors and retailers to purchase their raw agricultural commodities and to process and deliver them to the consumer for final sale. The result is the food and fiber system.

Within the agriculture industry, agribusiness is widely used simply as a convenient portmanteau of agriculture and business, referring to the range of activities and disciplines encompassed by modern food production. There are academic degrees in and departments of agribusiness, agribusiness trade associations, agribusiness publications, and so forth, worldwide. Here, the term is only descriptive, and is synonymous in the broadest sense with food industry. The UN's Food and Agriculture Organization, for example, operates a section devoted to Agribusiness Development , which seeks to promote food industry growth in the Third World.

Changing taste of consumers

Today the business has become very competitive and complex. This is mainly due to changing taste and fashion of the consumers on the one hand, and introduction of substitute and cheaper and better competitive goods, on the other.

The old dictum "produce and sells has changed overtime into "produce only what customers want". In fact, knowing what customers want in never simple.

Nevertheless, a farmer operator/farmer manager has to give proper thought to this consideration in order to make his business a successful one. The important requisites for success in a modern business are:

1. Determination of objectives: It is one of the most essential pre requisite for the success of business. The objectives set forth should be realistic and clearly defined. Then, all the business efforts should be geared to achieve the set objectives. In a way, objectives are destination points for an agribusiness. As a car driver must know here he/she has to reach, i.e. destination similarly business also must know what objectives.
2. Proper Planning: In simple words, planning is a pre-determined line of action. The accomplishment of objectives set, to a great extent, depends upon planning itself. It is said that it does not take time to do thing but it takes time to decide what and how to do. Planning is a proposal based on part experience and present trends for future actions. In other words, it is an analysis of a problem and finding out the solutions to solve them with reference to the objective of the farm.
3. Symmetrical organization: An organization is the art or science of building up systematical whole by a number of but related parts. Just as car is build up by various parts like engine, tyre, gear box etc. similarly, organization of business is a harmonies combination of men, machine material, money management etc. so that all these could work jointly as one unit, i.e. "business" "the agribusiness". Organization is, thus such a systematic combination of various related parts for achieving a defined objective in an effective manner.
4. Influence of Consumer Behaviour: As indicated earlier, today the agricultural production philosophy "produce what the consumer want". "Consumers" behaviour is influenced by variety of factors like cultural, social, personal and psychological factors. The business needs to know and appreciate these factors and then function accordingly. The knowledge of these factors is acquired through market research.

Research is a systematic search for new knowledge. Market research enable a business in finding out new methods of production, enhancing the quality of product and developing new products as per the changing tastes and wants if the consumers.

5. Finance: It is said to be the heart & soul of business enterprise. It combined by brings together the land, labour, machine and raw materials into production. Agribusiness should estimate its financial requirements adequately so that it may keep the business wheel on moving. Therefore, proper arrangements should be made for securing the required finance for the enterprise.
6. Right location, right place: The parameter of agribusiness success depends to mainly on the location. Where it is set up. Location of the business should be convenient from various points of view such as availability of required infrastructure facilities, availability of inputs like raw materials, skill labour, nearer to the market etc.

Hence the business men must take sufficient care in the initial stages to selected suitable location for his business.

The sine of the business is also important because the requirement for infrastructural facilities and inputs varies as per the size of the business. The requirement for raw materials, for example, will be les in a smaller sized firm than a larger size firm.

7. Efficient management: One of the reasons for failure of business often attributed to as their poor management or inefficient management. The one man, i.e. the proprietor may not be equally good in all areas of the business. Efficient businessman can make proper use of available resources for achieving the objectives set for the business.
8. Harmonious relations: In an agribusiness organization, the farmer operator occupies a distinct place because he/she is the main living factor among all factors of production.

In fact, it is the human factor who makes the use of other non-human factors like land, machine, money etc. Therefore, for successful operation of business, there should be cordial and harmonious relations maintained with the workers/labours to get their full cooperation in achieving business activities.

Scope of agribusiness

As it has been already mentioned that agribusiness is a complex, system of input sector, production sector, processing manufacturing sector and transport and marketing sector.

Therefore, it is directly related to industry, commence and trade, industry is concerned with the production of commodities and materials while commerce and trade are concerned with their distribution.

Industry

Industry refers to the processes of extraction and production of goods meant for final consumption or use buy individual or buy another industry for its production. Thus, goods used by the final or ultimate consumers are called "consumer goods" such as edible oils, fruit jams, papaya, pickles etc.

Types of Industries

As per the agribusiness concerned, the industries are broadly classified into following types.

(1) Extractive industries: These industries are concerned with the extraction; and utilization of natural resources. Example are fishing, fruit gathering, agro-based industries, forestation.

(2) Genetic Industries: These industries include breeding of plants, seeds, cattle breeding farm, fish hatcheries, poultry farms. Of course, factors like nature, climate and environment play a dominant role in these industries, yet human skill involved in their production cannot be ignored. For example intensive agriculture is possible with greater amount of capital and larger number of workers.

(3) Manufacturing Industries: These industries are engaged in the conversion of raw material or semi-finished goods produced in the extractive industries. Some prominent examples are – cotton textile industry, spinning and weaving mills etc. Manufacturing industries can further be classified into five types : (i) Analytical industry (ii) Processing industry (iii) Synthetic industry, (iv) Service industry (v) Assembly industry.

COMMERCE: MAJOR COMPONENT OF AGRIBUSINESS

Commerce is the another major component of agribusiness. It includes all those activities which are necessary to bring goods and services from the place of their production to the place of their consumption. Thus, sit includes the buying and selling of goods and service and all those activities which facilitate trade such as storing, grading, packaging, financing, insurance and transportation.

In simple words, commerce includes trade and aid to trade. The principal function of trade (commerce) are to remove the hindrance of person, place, time exchange, knowledge etc. and ensure a free and smooth flow of goods from the producers to the consumers.

Trade in fact is a branch of commerce itself. In a way, it is the final state of business activity involving sale and purchase of commodities or goods. It does not include and to trade like transportation, insurance, banking, finance etc. On the basis of its coverage and volume, trade is normally classified into the following types :

1. On the basis of volume

(i) Wholesale trade

(ii) Retail trade

2. On the basis of coverage

(i) Regional trade

(ii) National trade

Agribusiness is a broad concept used to describe corporate agricultural enterprises individually and collectively. Agribusinesses are companies involved in one or more stages of the production of crops and livestock. Examples of agribusiness activities include:

- Manufacture or distribution of agricultural supplies and equipment such as machinery, feed, and fertilizers
- Ownership or management of agricultural production facilities such as farmlands and livestock facilities
- Research and development of new agricultural resources and methods
- Processing or distribution of agricultural products

Providing food or fibers is the ultimate product of all agribusiness operations. As such, the economic impact of agribusiness is significant; agribusiness is almost two times as large as the sum of all manufacturing enterprises (measured in total assets); it represents 40 percent of all consumer spending; and it employs 37 percent of the labor force.

Agribusiness approach

The agribusiness approach is a method of examining farming problems in a new and more comprehensive setting. One benefit from this approach has been the release of workers—farm manpower—from agriculture for employment in new non-farm occupations—including the armed forces during wars. This has resulted in tremendous economic growth and development and an improved standard of living.

In the early nineteenth century, agriculture was a self-contained industry. The typical farm family produced its own food, fuel, shelter, draft animals, feed, tools, implements, and even clothing. Only a few necessities had to be bartered for or purchased off the farm. The farm family performed virtually all operations pertaining to the production, processing, storage, and distribution of farm commodities.

In the ensuing years, however, agriculture evolved from self-sufficiency to intricate interdependence with other segments of the economy, particularly those relating to the manufacture of production supplies and the processing and distribution of food and fiber products.

Agribusiness comparises of several million farm units and several thousand business units, each an independent entity, free to make its own decisions. Agribusiness is the sum total of hundreds of trade associations, commodity organizations, farm organizations, quasi-research bodies, conference bodies, and committees, each concentrating on its own interests. The U.S. government also is a part of agribusiness to the degree that it is involved in research, the regulation of food and fiber operations, and the ownership and trading of farm commodities. Land-grant colleges, with their teaching, experiment stations, and extension functions, form another sector of agribusiness. Agribusiness exists in a vast mosaic of decentralized entities, functions, and operations relating to food and fiber.

Use of New Technologies

One prominent feature of agribusiness is its continuous pursuit of new technologies. Well-known examples include the use of satellite-based global positioning systems (GPS) to closely manage crop lands and computer systems to manage various parts of the business. These technologies boost agricultural efficiency by reducing wasted resources, saving time, and improving output.

Crop agriculture, in particular, has turned to such high-tech solutions to develop what is known as precision or site-specific farming methods. These methods involve systematically testing crop fields for variations in fertility and soil composition. The data are then stored in a computer, and, using GPS equipment on the farm machinery, the on-board computer can then determine where in the fields to allocate seed and fertilizer to maximize yield and minimize waste.

AGRIBUSINESS IN INDIA

As incomes in India rise, tastes are changing. Staples are receding in importance. In 1983 food grains accounted for 32% of India's agricultural output; 15 years later they accounted for less than a quarter. But food grains still occupy nearly 64% of India's farmland and an even higher fraction of government concern and energy. In India, cereals are being increasingly overtaken by fruits, vegetables, milk, eggs and poultry.

India is the third-biggest producer of potatoes in the world. The humble spud finds itself stuffed into flatbread, encrusted in cumin seeds or tucked into pancakes. But the truckloads of large, oblong potatoes that arrive at the McCain Foods plant in the Mehsana district of Gujarat face a more exacting ordeal. Ferried by a conveyor belt and propelled by water, they are sized, steam-peeled, sliced, diced, blanched, dried, fried (for precisely 42 seconds in vegetable oil at 199°C), chilled, frozen, bagged and then boxed.

The fifteen kg boxes of fries that emerge at the other end of this pipeline supply the growing chain of McDonald's restaurants in India. When McDonald's first entered India in 1996, the food-processing industry was confined largely to ice-cream and ketchup. Even importing frozen fries was complicated by the fact that such an exotic item did not appear on India's schedule of tariffs and quotas. It took McDonald's roughly six years and $100m to weld a reliable supply chain together.

For fries, that supply chain begins with 2,000 acres of potato fields in Gujarat, cultivated by 400 farmers under contract with McCain Foods. These cultivators belong to a profession which still employs about half of India's workforce. In the home state of Mahatma Gandhi, agriculture is growing almost as quickly as the rest of the Indian economy. Gandhiji used to say that agriculture is the backbone of Indian economy. But elsewhere, agriculture is said to be in crisis. The average size of farmers' landholdings is only about 1.3 hectares.

If their fields are irrigated at all, they are flooded wastefully, with water flowing down furrows on either side of the crop, taking valuable nitrogen with it. India produces more tractors than any other country, but many farmers still use bullocks instead. They sell their produce at controlled prices in government markets: marketplaces regulated by the state with the aim of protecting farmers from exploitation by unscrupulous traders. State governments once took it upon themselves to spread know-how, market intelligence and the fruits of agricultural research to smallholders. But the agricultural extension system is now in some disrepair. Public investment in agriculture has stagnated over the past few years as the government's subsidy bill for food, fertiliser and fuel has risen.

In the dearth of public investment, Indian agriculture is increasingly dependent on private outlays, which now account for three-quarters of total investment in the sector. McCain Foods, for example, invested $25m in the Mehsana plant. And in the absence of government extension services, some private companies are finding alternative ways to let farmers know what the customer wants, and how to fulfil that expectation.

Hitesh Patel, for example, used to grow cottonseed on his six hectares in Idar village, about 125km from the Mehsana plant. Some years ago he planted a hectare of potatoes at McCain Foods' urging and under its guidance. Now, he plants potatoes on all of the six hectares he owns and another 1.6 he has leased.

McCain Foods offers him an assured price of 6.50 rupees ($0.14) per kilogram of potatoes, a better rate than the it. But they will not buy just any potato. India's common varieties are too small, watery and sugary, caramelising when fried.

Potatoes fit for processing are usually grown in more temperate latitudes. The tuber likes warm days and cool nights. "Wherever are the best regions for wine, potato is not far behind," says Ghislain Pelletier, a McCain agronomist. Firms before McCain had tried and failed to produce potatoes in India that were suitable for processing into fries. Even Harrison McCain, one of the company's founders, doubted it could be done.

But after several years of experimentation, McCain can now supply all of McDonald's needs in India, as well as producing some creations of its own, like Masala Fries. "We used to struggle for size; now we struggle to reduce the size," says Devendra Kumar of McCain India.

McCain's agronomists would first visit farmers every other day. Now they check in once a week. They insist that farmers give up flooding their furrows with water in favour of drip irrigation, which flows through a pipe punctuated with small nozzles laid along the crop-bed.

Drip irrigation produces "more crop per drop", as Mr Pelletier puts it. It moistens the soil at each root, but leaves the ground otherwise dry. This in turn reduces the humidity that attracts pests and blight.

This kind of contract farming began with Punjab farmers growing tomatoes for Pepsi's food business in the 1980s. But its spread was hampered by tight regulations at the state level on who could buy produce and how. Most states have now eased those restrictions, raising hopes that contract farming—which only accounts for a small fraction of India's annual agricultural output of $220 billion—will flourish.

McDonald's, for example, recently introduced the Chicken McNugget to India, where it counts as a premium product: all white meat without any bone. When the firm first entered the country, it struggled to find poultry suppliers who could debone meat. Because the chain's products are delicate or perishable they must be handled with far more care than rice or wheat. Frozen fries, for example, are so brittle that they must be handled like eggs. But McDonald's also found that the few refrigerated lorries in the country were mostly devoted to transporting ice-cream.

Investments by multinationals like McDonald's can spread into the wider economy. McDonald's, for example, invited East Balt Commissary of Chicago, whose founder supplied buns to Ray Kroc's first McDonald's franchise, to go to India to train the Cremica bakeries in Delhi and Mumbai. East Balt is what Abhijit Upadhye of McDonald's India calls a "good system player".

It also asked Schreiber International, which supplies sliced cheese, to tie up with Dynamix Dairy in India. As a result, many other companies, including Nestlé, Unilever and Pepsi, started doing brisk business with Dynamix.

Moreover, scores of research and development projects are conducted in agribusiness to find new technologies and to better use existing ones. Important R&D work includes developing genetically engineered crops, improving the pest resistance of crops, using biotechnology in agriculture, and formulating new agricultural pharmaceuticals and chemicals. Despite considerable government and university backing for agribusiness research, the majority of this research is funded by the private sector.

The Global Agri-Food System

Technological advancement in twenty first century has given new dimension to agri-food system.

In an era of rapid change and growing risk and uncertainty, agricultural policy and practice in the developing world is encountering, a number of limitations which reveal inadequacies in its long-term sustainability and its capacity to meet the range of objectives that it is expected to deliver. These include concerns about chronic hunger and malnutrition, adverse environmental changes, the limits of technology-enhanced productivity gains, increasing land exploitation and the loss of biodiversity, livelihood insecurity and the continuing poverty of agricultural communities. Worries about food safety, hygiene and nutrition, and growing demands for the re-localisation of agri-food systems from citizen consumers in both the North and the South have also emerged. These apprehensions raise important questions about whether the forms of agriculture developed over the past century, and celebrated as technically advanced and 'modern', are able to respond to the complex and diverse array of in this new era.

FEATURES OF AGRI-FOOD SYSTEMS

Agriculture therefore has a number of features that distinguish it from other productive sectors. Among other things, it is mainly a private activity implemented locally mostly by households, but it also has many dimensions of collective action, is deeply affected by global forces and depends greatly on public interventions for its structure, its support and its development. Taking into account these drivers of change, what are the key characteristics of agri-food systems that demand attention? The following, we suggest, are critical:

- The dynamics of production. Agriculture is characterised by high dependency on natural resources, spatial dispersion of activity, seasonal variables, asymmetries in information due to location and distance, high risks associated with the vagaries of nature, and difficulties in sustaining the productivity of natural resources, because their use and reproduction typically conflict.
- Integrated agri-food systems. Market developments, technological progress, institutional changes and policy interventions in one part of the world have far-reaching implications, even for distant actors, as global and regional supply chains link producers and consumers in different parts of the world.
- Market failures. Failures of input and output markets for agricultural goods and services are associated with high transactions costs, particularly adversely affecting poor farmers, information asymmetries, incomplete property rights, externalities and missing actors. These affect access to markets, the availability of insurance and financial services, and the underwriting of contracts.
- Public sector interventions. There is a signiûcant need for public sector interventions to compensate for these market failures. This makes agriculture highly vulnerable to extractive policies (cheap food policies), land grabbing, rent seeking, regressive subsidies and exposure to corrupt officials. Public budgets in agriculture can easily be prey to clientelism and elite capture, major causes for the mis-investment of already undersized public budgets.
- Equally, they can be distorted and misdirected through the changing whims and misguided policy prescriptions of international donors. Consequently, the political economy of policy and investment in agriculture can determine success or failure in agricultural development.
- Socio-cultural systems. The close correspondence between agriculture as a productive activity ('agri-') and rurality as a way of life ('-culture') make social relations in rural society important determinants of access to resources (for example through land rental markets), asymmetries in power (including by gender and ethnicity) and benefits from public services. These affect dynamic poverty outcomes in agriculture.

- Heterogeneity and diversity. Actors with better asset endowments in favourable areas can take advantage of new markets and of new technological and institutional opportunities. In contrast, large segments of the smallholder population remain reliant on subsistence-oriented activities, linked to labour markets as net sellers and to food markets as net buyers, but relying on agriculture for home consumption and as a safety net of last resort.
- Collective action. Eûective forms of cooperation and collective action are essential for the millions of smallholders, pastoralists, fisher folk and farmworkers to have their voices heard in key regional and national policy forums. They are also essential to enable producers' associations and federations achieve economies of scale (meeting new market requirements (e.g. grades and standards) and interacting in local clusters of economic activity), access public services and manage common property resources.

The key question – addressed in a preliminary way by this paper and the wider work on food and agriculture by the STEPS Centre – is: how, in the face of these old and new challenges, can poor, marginal people negotiate pathways to sustainability through agriculture? This paper therefore provides an assessment of how different visions of agricultural development, underpinned by contrasting narratives of technological and economic change, respond to these challenges. Yet, despite the diversity of drivers and contexts, the complexity of dynamics and the uncertainties that prevail, current debates about agricultural change in developing countries are often couched – implicitly, if not explicitly – in terms of notions of 'progress' towards a singular goal.

Frequently, the underlying assumption is that such progress is achieved through the transfer of knowledge, ideas, models, practices and technologies from the 'developed' world to the 'developing' world, or from 'modern' science to 'backward' farming settings. For example, in the early 1990s some Indian agricultural universities based their curriculum on two decade old textbooks from the United States, as if no new, locally-speciûc innovations had happened in the intervening years. Within this framing are often embedded notions about how agricultural development occurs in a linear sequence of stages – from 'backward' to 'modern', from 'old' to 'new', from 'underdeveloped' to progressively more 'developed', from pre-industrial to industrial. Thus, there is often assumed to be a singular path to progress, and to be committed to this path governments, farmers, aid agencies and analysis have to be uniformly and unquestioningly 'pro-innovation', 'pro-technology' and 'proudevelopment'.

Those who criticise this monolithic linear assumption are sometimes accused of being 'anti-technology' or 'anti-modernity'. Our approach is not hostile to beneûcial technological changes, but we do reject approaches that assume that there is one, and only one, technological trajectory that implicitly denies the existence and benefits of alternative pathways, or even multiple pathways, towards a broader goal of poverty reduction, social justice and environmental sustainability.

Thus, in considering possible pathways towards this goal, we assume that there may be multiple routes to improving the relationship between complex food system and poor people in developing countries, and that poor rural people often have relevant agricultural knowledge.

Agricultural Sustainability

The term "agribusiness" was coined in the 1950s by John Herbert Davis and Ray A. Goldberg to reflect the two-way interdependence between businesspeople and farmers in the dual roles of suppliers and purchasers. Business firms that serve agriculture rely on farmers for their markets and for some of their supplies. By the same token, farms could not operate without businesses that manufacture farm supplies and those that store, process, and merchandise farm commodities.

Over the past 20 years, a great deal of work on agricultural sustainability has focused on the capacity of food systems to absorb perturbations and still maintain their functions. In a resilient and robust agri-food system, disturbances have the potential to create opportunities for doing new things, for innovation and for new pathways of development. In a vulnerable system, even small disturbances may cause signiûcant adverse social consequences especially for those who are most vulnerable, such as the rural poor in develop ing countries. Until recently, dominant perspectives in conventional agricultural science and development programmes have implicitly assumed a stable and almost indeûnitely resilient environment, where resource flows could be controlled and nature would return to a steady state when human stressors were removed.

Such static, equilibrium-centred views, we argue, provide inadequate insight into the dynamic character of agri-food systems, particularly in an era of global economic and environmental change, where factors such as climate change, rapid land use shifts and uncertain political economic conditions in agricultural economies all impinge on the day-today realities of poorer producers and consumers in the developing world.

Our focus on uncertainty, complexity and diversity is at the core of the STEPS Centre agenda, and aims to shift attention from policies and practices that aspire to maintain the status quo or control change in systems assumed to be stable, in favour of analysis and practices that enhance the capacity of agri-food systems to respond to, cope with and shape change. Such responses in turn enhance the possibilities of sustaining desirable, yet diverse, pathways for development in changing environments, where the future is unpredictable and surprises are likely.

Agricultural & Resource Management Problems

Much conventional agricultural science and policy does not seem to be able to explain, let alone respond to, complexity, diversity, uncertainty and non-equilibrium states, although poor people who are dependent on agriculture for their livelihoods very often live in complex, diverse and risk-prone settings, with inherent seasonal instability.

Vulnerability not only damages people's welfare, it also reduces growth, both directly by destroying assets and indirectly as threats of shocks and stresses cause assets to be diverted assets from more productive activities to those that reduce risk and uncertainty.

Agricultural and resource management problems typically tend to be classic 'systems' problems, where aspects of systems behaviour are both complex and unpredictable and where causes, while at times apparently simple, when ûnally understood are always multiple. These problems are often non-linear in nature, cross-scale in time and space and dynamic in character. This is true for both natural and social systems and their interactions. In fact, they need to be understood as one system, with critical feedbacks across temporal and spatial scales. Thus, interdisciplinary and integrated modes of inquiry are needed for understanding and designing eûective responses to human–environment interactions related to food and agriculture in a turbulent world. A critical minority of policy-makers and citizens – both producers and consumers are demanding integrated solutions that address these issues of uncertainty, iversity and complexity. Their calls for action are not so much driven by prophesies of doom as by the need for understanding and action. But if you seek understanding, to whom do you turn for information and advice?

Agricultural science often provides only limited assistance, largely because it includes not only conflicting voices – witness the debate on genetically modified (GM) crops or arguments over food production vs. population growth – but also conflicting modes of inquiry and criteria for establishing the trustworthiness of different lines of argument.

In particular, the philosophies of two streams of agricultural science are often in opposition. The tension between them is now evident in biology. One stream is represented by the paradigm of molecular biology and genetic engineering. This stream of science promises to provide not only health and economic benefits from agricultural biotechnology, but also an uncertain era of changing social values and consequences.

This stream is a science of parts; that is an analysis of speciûc biophysical processes that affect survival, growth and distribution of target variables as if they were independent of each other and could be systematically controlled one at a time. It emerged from a tradition of experimental science, where a narrow enough focus is chosen to pose speciûc questions and empirical hypotheses, collect data and design critical tests for the rejection of falsiûed hypotheses. The goal is to narrow uncertainty to the point where acceptance of an argument among scientific peers is essentially unanimous. Thus, it is conservative and narrowly focused, and it achieves this by being fragmentary and incomplete. It provides individual building blocks of an ediûce, but not the architectural design. This kind of approach to modern agricultural science, a science of the parts, may be suitable for certain types of conventional agricultural development but not for sustainable agriculture – if sustainable agriculture is deûned more broadly to include a range of ecological, economic and social objectives, such as sustained reductions in chronic malnutrition, poverty and ecological harm.

By contrast, a holistic stream can be characterised as a science of integration; that is, by inter-disciplinarity and synthesis, by cross-sectoral and cross-scale research and analyses. It is represented, for example, by agroecology, conservation biology, landscape ecology and other systems approaches that include the analysis of (agri-food)-ecosystems, the interactions between multiple coexisting populations and landscapes, and more recently, the study of socio-ecological dynamics at different scales and concerns about global environmental change, such as climate change.

The applied forms of this stream have emerged regionally in new forms of integrated agricultural practice and natural resource and environmental management, where uncertainty and surprises become an essential part of an anticipated set of adaptive responses. They are fundamentally about blending disciplinary perspectives and combining historical, comparative and experimental approaches at scales appropriate to the issues. It is a stream of investigation that is fundamentally concerned with integrative modes of inquiry and multiple sources of evidence.

SEARCH FOR SUSTAINABLILITY SOLUTIONS

Characteristics of Agri-food Systems

If a consideration of dynamic uncertainty needs to be at the core of any search for sustainable solutions to developing world agriculture, what, then, are the factors that drive change and create risks and uncertainties in developing world agriculture today? This section explores this question through an assessment of key drivers of change and their eûects that characterise contemporary agrifood systems.

Agriculture is an important source of livelihoods in developing countries, providing ways of life for billions of people, many of them poor. Of the world's 6.5 billion inhabitants, 5.5 billion live in developing countries, 3 billion in the rural areas of these countries (World Bank 2007).

Of rural inhabitants, an estimated 2.5 billion are involved in agriculture, 1.3 billion are smallholders, while others include farm labourers, migrant workers, herders, fishers, artisans and indigenous peoples who depend on agriculture and natural resources for their livelihoods. More than half are women. The developing world will remain predominantly rural until around 2020 and millions of poor people in those countries will continue to rely on agriculture for their livelihoods for the foreseeable future.

The contribution of agriculture to livelihoods is evident from the fact that 70 percent of the world's poor people, including the poorest of the poor, and 75 percent of the world's malnourished live in rural areas, where most of them are involved in agriculture. The Millennium Development Goal of halving extreme poverty and hunger will not be met without reducing this rural poverty (UN Millennium Project 2005a).

Meeting this food security goal will be a major challenge. Yet rural poverty remains stubbornly high, even with rapid growth in the rest of the economy. Rural-urban income gaps tend to rise as non-agricultural growth accelerates, creating major social tensions as expectations for better lives remain unfulûlled for a majority of the rural people.

Given these trends, chronic hunger and global food security will remain a worldwide concern for the next fifty years and beyond, as the world's population grows from its current 6.5 billion to upwards of ten billion, most of whom will reside in developing countries. Of course, predictions of food security outcomes have been a part of the policy discourse in agriculture at least since the Reverend

Thomas Malthus wrote An Essay on the Principle of Population in 1798. Over the past several decades, some neo-Malthusians or 'catastrophists' have expressed concern about the ability of agricultural production to keep pace with global food demands whereas other 'cornucopians' have forecast that technological advances or expansions of cultivated area would boost production suûciently to meet rising demands. Thus far, dire predictions of a global food security catastrophe have proved unfounded, in the sense that ag- gregate food supply has kept pace with population growth, although hundreds of millions remain hungry and malnourished.

Nevertheless, despite the fact that food production per capita has been increasing globally, major distributional inequalities remain, linked primarily to poverty. Global food production has increased by well over 130 percent since the 1960s, yet the fact that almost 78 per cent of countries that report child malnutrition are food-exporting countries dramatically illustrates a 'paradox of plenty'. Moreover, the productivity of major cereals appears to be reaching biological limits in some regions, despite heavy use of agrochemical inputs, and consequently production is now growing more slowly than in recent decades. Widespread and persistent hunger is a fundamental contradiction in today's world when production and productivity in agriculture have grown faster than eûective demand. An estimated 852 million people were undernourished in 2000–02, up 37 million from the period 1997–99. Of this total, more than 95 percent live in developing countries (World Bank 2006). Sub-Saharan Africa, the region with the largest share of undernourished people, is also the place where per capita food production has lagged the most.

This underperformance of the agriculture sector has been exacerbated by ethnic conflict and political instability, declining terms of trade, dwindling investments in agricultural research and infrastructure and increasing water scarcity.

The challenges faced today are, however, substantially different from those encountered by the green revolution producers who achieved sustained gains in agriculture productivity only a few decades ago. Since the 1980s, there has been a substantial decline in public sector support for agriculture and many producers have lost access to key inputs and services. While public sector provision of these services was never very eûcient, it often provided the linkages to markets for poor rural producers.

Today, such links are tenuous and complicated by much greater integration of the global economy. Smallholder producers now compete in global markets that are much more demanding in terms of grades and standards (e.g. quality, traceability and food safety), and more concentrated and vertically integrated than in the past.

A major concern about this concentration is the control exercised by a handful of private corporations over decision-making throughout the agri-food system. In the past, most of the global trading and grain-handling farms were family-held operations which operated in one or two stages of the food system and in a very few commodities. Consequently, risk exposure of many rural households to market forces was very different from that today. Risks were often reduced by the state through government-controlled marketing boards and similar parastatal organisations, which assured a price structure, input and output markets and access to improved technologies and training. Public investments in research and development resulted in higher yielding farm systems.

Furthermore, innovations were encouraged through public subsidies of one kind or another. In much of Asia and Latin America these innovations led many farm households to shift to more productive and higher return farming systems (World Bank 2005).

Today, the system is becoming much more complex, starting with a farm's involvement in (bio) technology, extending through agro-chemical inputs and production, and ending with highly processed food. Increasingly, these farms are developing a variety of different alliances with other players in the system, forming new food system 'clusters'.

As agriculture becomes more concentrated and integrated, these giant clusters are establishing an oligopsony – a market in which a small number of buyers exerts power over a large number of sellers – over large parts of the agri-food chain, enabling them to maximise profit while minimising risk. As a result, the food system has begun to resemble an hourglass.

At the bottom are millions of farmers and farm labourers producing the food and fibre, while at the top are billions of consumers, both rich and poor. At the narrow point in the middle are the dozen or so multinational corporations – the input suppliers, food processors and retailers – earning a profit from every transaction. Typically, goods are exchanged through closed contracts or intra-farm transfers rather than open wholesale markets and even when they are exchanged in wholesale markets, prices may be well below the cost of production due to oversupply. Consequently, the 'cost-price squeeze' falls on the producers, who bear the bulk of the risk and share little, if any, of the rewards.

Because agriculture has a larger tradable component than most sectors, it is profoundly affected by the trade environment and trade policy. Whereas overall trade barriers in industrial countries have declined signiûcantly over the last decade, the remaining barriers are concentrated on agricultural products and labour-intensive manufactures in which developing countries have a comparative advantage.

High levels of farm support, at the level of USD $279 billion (EUR €226 billion) per year in countries belonging to the OECD, depress world prices for several key commodities (especially sugar, cotton, milk, and beef) and deeply undermine agricultural growth in developing countries. Quotas and tariffs remain important instruments for protection, and sanitary and phytosanitary restrictions increasingly perform the same function.

Global and regional economic integration is accompanied by other challenges that further weaken the socioeconomic position of the rural poor. In some parts of the world, especially in Sub-Saharan Africa, rural areas are hard hit by the HIV/AIDS pandemic, which is disrupting the transfer of knowledge, destroying traditional land allocation systems, and dramatically changing the demographic composition of many rural communities.

Global environmental change is increasing pressure on an already fragile natural resource base in complex, risk-prone environments that are the mainstay of rural livelihoods. Rising energy prices are driving massive investments in biofuels, which could increase the volatility of food prices with negative food security implications in some regions (UN-Energy 2007). Finally, conflict conditions, many of which result from or are provoked by poverty, are further eroding the livelihood systems and resilience of poor rural people.

This is not helped by the fact that attention to agriculture in terms of policy commitments and investment levels declined in both international donor and developing country policies and programmes, despite the demonstrated high rates of return and the reductions in poverty that come from such investments. Further progress is curtailed by weaknesses and deficiencies in agricultural science and technology policy regimes that result in institutional arrangements and organisational forms unsuited to development and broad-based diûusion of poverty-reducing innovations. Investments in science and technology have been shown to pay oû most strongly for countries and regions with highly integrated technical and economic systems able to diffuse and apply results of new research.

Constraints faced in mobilising public resources for agricultural development in countries with widespread poverty and undernourishment are illustrated by relating government expenditure on agriculture to the size of the agricultural labour force. In countries where more than 35 percent of the people are undernourished, government expenditure per agricultural worker averages USD $14 or 50 times less than the USD $880 in countries with the lowest rates of undernourishment.

Similar problems also affect agricultural R&D. Corporate R&D agendas understandably focus on potentially profitable sectors, which frequently do not include poor people. Only in the public and charitable sectors have agricultural research and development policies engaged with the needs of the rural poor.

Assumptions about the vulnerability and/or robustness and resilience of gricultural systems remain contested. Many studies predict that world food supplies may not necessarily be adversely affected by moderate climate change, but only by assuming farmers will take adequate steps to adapt to climate change and that in some regions additional CO2 will contribute positively to increased yields. Many developing countries are likely to fare badly, however, as climate change may result in unpredictable growing conditions, including more intense rainfall events between prolonged dry periods, as well as reduced or more variable water resources for irrigation in tropical environments.

Increasing agricultural expansion into marginal lands and forests may in turn put these areas at greater risk of environmental degradation. Such conditions may promote pests and disease on crops and livestock and increase the incidence of vector-borne diseases in humans, as well as increase soil erosion and desertiûcation.

The HIV/AIDS pandemic is another relevant concern for the sustainability of agriculture. More than 28 million people have died since the first case was reported in 1981. In 2005, AIDS killed 2.8 million people, and an estimated 4.1 million became infected, bringing to 38.6 million the number of people living with the virus around the world. Of this total, 24.5 million of these people live in SubSaharan Africa (where in some countries one in three adults are infected) and 8.3 million live in Asia. In addition to its direct health, economic and social impacts, the disease also affects food security and nutrition.

Adult labour is often reduced or removed entirely from affected households, and those households then have less capacity to produce or buy food, as assets are often depleted for medical and/or funeral costs. The agricultural knowledge-base often deteriorates as individuals with farming experience and scientific knowledge succumb to the disease.

Moreover, agri-food systems are changing in many ways as a result of the dynamic interactions of a range of environmental and socioeconomic drivers, including global environmental change, agricultural intensification, concentration of production, vertical integration and coordination, industrialisation, deregulation and economic liberalisation and urbanisation.

Essential Task of Management

If the manager fails to capitalize on the energies supporting these employee goals, those same energies are sure to become the business's loss. Management scholar Douglas McGregor argues that, "the essential task of management is to arrange organizational conditions and methods of operation so that employees can achieve their own goals best by directing their own efforts towards those objectives set by the organization." The managers of the "old school" might argue that such a permissive environment will only result in diminished productivity.

But allowing for the fact that all managers must, at times, play a dictatorial role, it would seem that employees would respond more positively to such moments of toughness, if it is encompassed within a past record of individual respect, consideration and fair play.

In conclusion, a rather simple rule emerges, i.e., the respectful treatment of others is likely to be reciprocated. And it is this reciprocation that transforms a manager into a leader. Lao-Tsi was a Chinese poet and philosopher who lived 2,500 years ago. He likely possessed little experience as a manager, yet his words of advice are still timely and relevant.

He wrote, "Fail to honor people, and they will fail to honor you; but of a good leader, who talks little, when his work is done, his aim fulfilled, they will say: we did this ourselves."

Agribusiness System

The term value chain was first popularized in a book published in 1985 by Michael Porter, who used it to illustrate how companies could achieve what he called "competitive advantage" by adding value within their organization. Subsequently, the term was adopted for agricultural development purposes and has now become very much in vogue among those working in this field, with an increasing number of bilateral and multilateral aid organisations using it to guide their development interventions.

At the heart of the agricultural value chain concept is the idea of actors connected along a chain producing and delivering goods to consumers through a sequence of activities. However, this "vertical" chain cannot function in isolation and an important aspect of the value chain approach is that it also considers "horizontal" impacts on the chain, such as input and finance provision, extension support and the general enabling environment. The approach has been found useful, particularly by donors, in that it has resulted in a consideration of all those factors impacting on the ability of farmers to access markets profitably, leading to a broader range of chain interventions. It is used both for upgrading existing chains and for donors to identify market opportunities for small farmers.

Inputs Sector

Agricultural supplies

An agricultural supply store or agrocenter is an agriculturally-oriented shop where one sells agricultural supplies—inputs required for agricultural production such as pesticides, feed and fertilizers . Sometimes these stores are organized as cooperatives, where store customers aggregate their resources to purchase agricultural inputs. Agricultural supply and the stores that provide it are part of the larger Agribusiness industry.

Agricultural labor

A farmworker or agricultural worker is someone employed for labor in agriculture. In labor law, the term "farmworker" is sometimes used more narrowly, applying only to a hired worker involved in agricultural production, including harvesting, but not to a worker in other on-farm jobs, such as picking fruit.

Agricultural work varies widely depending on context, degree of mechanization and crop. In countries like the United States where there is a declining population of American citizens working on farms—temporary or itinerant skilled labor from outside the country is recruited for labor intensive crops like vegetables and fruits.

Agricultural labor is often the first community effected by the human health impacts of environmental issues related to agriculture, such as health effects of pesticides or exposure to other health challenges such as valley fever. To address these environmental concerns, immigration challenges and marginal working conditions, many labor rights, economic justice and environmental justice movements have been organized or supported by farmworkers.

Seeds

Seed companies produce and sell seeds for flowers, fruits and vegetables to commercial growers and amateur gardeners. The production of seed is a multibillion-dollar business, which uses growing facilities and growing locations worldwide. While most of the seed is produced by large specialist growers, large amounts are also produced by small growers that produce only one to a few crop types. The larger companies supply seed both to commercial resellers and wholesalers. The resellers and wholesalers sell to vegetable and fruit growers, and to companies who package seed into packets and sell them on to the amateur gardener.

Most seed companies or resellers that sell to retail produce a catalog, for seed to be sown the following spring, that is generally published during early winter. These catalogs are eagerly awaited by the amateur gardener, as during winter months there is little that can be done in the garden so this time can be spent planning the following year's gardening. The largest collection of nursery and seed trade catalogs in the U.S. is held at the National Agricultural Library where the earliest catalogs date from the late 18th century, with most published from the 1890s to the present.

Seed companies produce a huge range of seeds from highly developed F1 hybrids to open pollinated wild species. They have extensive research facilities to produce plants with genetic materials that result in improved uniformity and appeal. These qualities might include disease resistance, higher yields, dwarf habit and vibrant or new colors. These improvements are often closely guarded to protect them from being utilized by other producers, thus plant cultivars are often sold under the company's own name and protected by international laws from being grown for seed production by others. Along with the growth in the allotment movement, and the increasing popularity of gardening, there have emerged many small independent seed companies. Many of these are active in seed conservation and encouraging diversity. They often offer organic and open pollinated varieties of seeds as opposed to hybrids. Many of these varieties are heirloom varieties. The use of old varieties maintains diversity in the horticultural gene pool. It may be more appropriate for amateur gardeners to use older (heirloom) varieties as the modern seed types are often the same as those grown by commercial producers, and so characteristics which are useful to them (e.g. vegetables ripening at the same time) may be unsuited to home growing.

Fertilizers

Fertilizer is any material of natural or synthetic origin that is applied to soil or to plant tissues to supply plant nutrients. Fertilizers may be distinct from liming materials or other non-nutrient soil amendments. Many sources of fertilizer exist, both natural and industrially produced. For most modern agricultural practices, fertilization focuses on three main macro nutrients: Nitrogen (N), Phosphorus (P), and Potassium (K) with occasional addition of supplements like rock dust for micronutrients. Farmers applying these fertilizers in a variety of ways: through dry or pelletized or liquid application processes, using large agricultural equipment or hand-tool methods.

Historically fertilization came from natural or organic sources: compost, animal manure, human manure, harvested minerals, crop rotations and byproducts of human-nature industries (i.e. fish processing waste, or bloodmeal from animal slaughter). However, starting in the 19th century, after innovations in plant nutrition, an agricultural industry developed around synthetically created fertilizers. This transition was important in transforming the global food system, allowing for larger-scale industrial agriculture with large crop yields. In particular, nitrogen-fixing chemical processes such as the Haber process at the beginning of the 20th century, amplified by production capacity created during World War II led to a boom in using nitrogen fertilizers. In the later half of the 20th century, increased use of nitrogen fertilizers (800% increase between 1961 and 2019) have been a crucial component of the increased productivity of conventional food systems (more than 30% per capita) as part of the so-called "Green Revolution".

Production Sector

Farming

A farm (also called an agricultural holding) is an area of land that is devoted primarily to agricultural processes with the primary objective of producing food and other crops; it is the basic facility in food production. The name is used for specialized units such as arable farms, vegetable farms, fruit farms, dairy, pig and poultry farms, and land used for the production of natural fiber, biofuel and other commodities. It includes ranches, feedlots, orchards, plantations and estates, smallholdings and hobby farms, and includes the farmhouse and agricultural buildings as well as the land. In modern times the term has been extended so as to include such industrial operations as wind farms and fish farms, both of which can operate on land or sea.

There are about 570 million farms in the world, with most of which are small and family-operated. Small farms with a land area of less than 2 hectares operate about 1% of the world's agricultural land, and family farms comprise about 75% of the world's agricultural land. Modern farms in developed countries are highly mechanized. In the United States, livestock may be raised on range, land and finished in feedlots and the mechanization of crop production has brought about a great decrease in the number of agricultural workers needed. In Europe, traditional family farms are giving way to larger production units. In Australia, some farms are very large because the land is unable to support a high stocking density of livestock because of climatic conditions. In less developed countries, small farms are the norm, and the majority of rural residents are subsistence farmers, feeding their families and selling any surplus products in the local market. Acres can hold the crops.

Farm Mechanization

Agricultural engineering, also known as agricultural and biosystems engineering, is the field of study and application of engineering science and designs principles for agriculture purposes, combining the various disciplines of mechanical, civil, electrical, food science, environmental, software, and chemical engineering to improve the efficiency of farms and agribusiness enterprises as well as to ensure sustainability of natural and renewable resources.

An agricultural engineer is an engineer with an agriculture background. Agricultural engineers make the engineering designs and plans in an agricultural project, usually in partnership with an agriculturist who is more proficient in farming and agricultural science.

Processing Sector

Primary Processing

Primary food processing turns agricultural products, such as raw wheat kernels or livestock, into something that can eventually be eaten. This category includes ingredients that are produced by ancient processes such as drying, threshing, winnowing and milling grain, shelling nuts, and butchering animals for meat. It also includes deboning and cutting meat, freezing and smoking fish and meat, extracting and filtering oils, canning food, preserving food through food irradiation, and candling eggs, as well as homogenizing and pasteurizing milk.

Contamination and spoilage problems in primary food processing can lead to significant public health threats, as the resulting foods are used so widely. However, many forms of processing contribute to improved food safety and longer shelf life before the food spoils. Commercial food processing uses control systems such as hazard analysis and critical control points (HACCP) and failure mode and effects analysis (FMEA) to reduce the risk of harm.

Secondary Processing

Secondary food processing is the everyday process of creating food from ingredients that are ready to use. Baking bread, regardless of whether it is made at home, in a small bakery, or in a large factory, is an example of secondary food processing. Fermenting fish and making wine, beer, and other alcoholic products are traditional forms of secondary food processing. Sausages are a common form of secondary processed meat, formed by comminution (grinding) of meat that has already undergone primary processing. Most of the secondary food processing methods known to human kind are commonly described as cooking methods.

Marketing Sector

Agricultural marketing covers the services involved in moving an agricultural product from the farm to the consumer. These services involve the planning, organizing, directing and handling of agricultural produce in such a way as to satisfy farmers, intermediaries and consumers. Numerous interconnected activities are involved in doing this, such as planning production, growing and harvesting, grading, packing and packaging, transport, storage, agro- and food processing, provision of market information, distribution, advertising and sale. Effectively, the term encompasses the entire range of supply chain operations for agricultural products, whether conducted through ad hoc sales or through a more integrated chain, such as one involving contract farming.

Farmers' Market

A farmers' market (or farmers market according to the AP stylebook, also farmer's market in the Cambridge Dictionary) is a physical retail marketplace intended to sell foods directly by farmers to consumers. Farmers' markets may be indoors or outdoors and typically consist of booths, tables or stands where farmers sell their produce, live animals and plants, and sometimes prepared foods and beverages. Farmers' markets exist in many countries worldwide and reflect the local culture and economy. The size of the market may be just a few stalls or it may be as large as several city blocks. Due to their nature, they tend to be less rigidly regulated than retail produce shops.

They are distinguished from public markets, which are generally housed in permanent structures, open year-round, and offer a variety of non-farmer/non-producer vendors, packaged foods and non-food products.

2

AGRIBUSINESS MANAGER: JOB DESCRIPTION

For decades, food security has been a worldwide problem affecting the poorest in order to find apossible solution to this dilemma that not only continues, but constantly grows with theincrease in population. Agriculture, as the main food supplier, has evolved and increased itsproductivity, and has expanded its agricultural frontiers, since previous decades, in its attemptsto solve this problem. If predominant conventional agriculture (mainly dedicated toproductivity) is complemented with substantial increase of transgenic, organic,agroecological, greenhouse and hydroponic crops, it will be more feasible to avoid the lo ss offorest, the damage and loss of soils and at the same time diminish food insecurity. Within thiscontext, the agribusiness manager and other professionals related with the farming andeconomic activities, have given an ample and important contribution.

From the management ofagriculture at local, national and international levels to the management of livestock and otherrelated enterprises, being the bases of food production in any country, agriculturaladministrators have substantially helped to solve the problem. Their contribution can also beseen in education, investigation, training and political design.

The Agribusiness Management (ABM) program focuses on system-wide business and managerial functions performed by organizations throughout the agribusiness sector.

The major gives students the flexibility to choose courses related to their individual interests in farm management, commodity procurement and marketing, labor and personnel management, public policy and global food issues.

A career as agricultural managers oversees farm operations. A farming manager job involves activities such as inspecting diseased plants and working with farm equipment, such as tractors and hay pickers. Work can be dangerous, and caution must be exercised when dealing with chemicals and heavy machinery for husbandry managers. At large farms, agricultural management jobs primarily handle administrative tasks, such as hiring or marketing. A strategy may be required for a career as agricultural managers while working with employees or customers to challenge.

An agribusiness manager is any other manager who oversees the whole business from the planning, to carrying out duties and responsibilities that are needed for the business to be a successful one. The main difference is that agribusiness managers work in farms and other agriculture companies. Agribusiness managers analyze and figure out marketing strategies, hire and mentor the workers of the farm and may also perform some of the farming duties themselves. They are also responsible in planning where and how much to sell the crops and livestock.

DUTIES

Agribusiness managers evaluate the business and the amount of money a farm or agricultural business is raking in. They keep records of planting and harvesting times and identify when a crop or livestock is in demand. They also look after the people working in the business. Agribusiness managers coordinate and give workshops in agricultural marketing and its issues that will help in straightening out the problems that the staff is experiencing. They also prepare and develop training modules. Agribusiness managers make a database that supports their establishment in order for the farmers as well as the consumers to be well informed of the stocks that they have or are offering. They sometimes put up a website as well as a marketing strategy. It is also easier for the agribusiness's clients to reach them and see what their business is all about. An agribusiness manager also deals and contacts exporters and logistics to be able to ship out the products being offered.

Education and Training Requirements

To become an agribusiness manager, one must acquire a master's degree in agricultural economics, business administration, marketing, or any other related field. One must have experience in agribusiness or agribusiness development. Experience in handling a farm, or work experience in economics is a big advantage.

Knowledge and Skills Requirements

To become an agribusiness manager, one has to have good communication skills, both written and oral. One must have experience in dealing with suppliers, traders and exporters. He/she must have experience in marketing and economics, whether it be for agribusiness or not. Experience in farming, crops and livestock is also needed.

Working Conditions

An agribusiness manager usually works the standard 40 hours a week in an office or on the field. His/her work is not limited to just paper work and contacting vendors, suppliers and exporters. In small-medium scale businesses, some agribusiness managers go on the field and look out for the crops being planted – some even plant the crops themselves.

Types of a Agricultural Manager

There are types of agricultural managers. We have mentioned below several major types of farming manager roles with varying job titles.

Farm Manager: Farm managers primarily look after the farms and all the other related agricultural activities. It is their duty to ensure that the farm is being run smoothly and that there are no hurdles on the way. Agriculture farm manager jobs are to plan out activities that will help in increasing the productivity of the farm without harming the environment. Management is a core responsibility of agricultural management jobs. They look after the workers as well and ensure that the work they do is up to the point.

Livestock Manager: As the name suggests, livestock managers are responsible to look after poultry, dairy farms. It is their essential duty to ensure that the animals living on the farm are well looked after and that they do not have any diseases. Apart from ensuring that the animals are safe, livestock managers often manage the entire farms and look after other logistic needs as well. they are often the point person at a farm and supervise all the activities.

Horticulture Manager: Horticulture managers are those who look after the agricultural plants in a controlled environment. These days there is a lot of focus on eating clean and healthy and this is where horticulture managers come into the picture. They help the farmers to grow fruits and vegetables and do it in such a way that there is a minimum environmental impact. They also carry out maintenance work of storage spaces where fruits and vegetables are stored. Agricultural management jobs are to work towards keeping the crop pest free.

WORK OF AGRIBUSINESS MANAGE

The agribusiness manager's job is similar to any other manager, the difference is that this business is agriculture and they most probably work on farms. They need to have in depth knowledge about farming, cattle and all other agriculture related businesses. They may not be working in a corporate like office, but they have to be just as analytical.

The agribusiness managers have to have knowledge about the latest market conditions. They need to have good contacts with buyers and suppliers. They have to get the best prices for all the products, because the food prices depend upon the agriculture business, and food prices affect inflation.

Planning and Implementation

The managers have to hire workers required on the farm. They have to provide training to them. They have to manage their shifts and decide their wages. The agribusiness is a seasonal business; hence, the manager has to plan accordingly. What crops to plant and in which season all has to be decided by the managers. The manager also has to arrange for the workers' employment in the off-season.

These tasks are all a part of the agribusiness manager job description. If cattle breeding is the main business the manger is looking after, they have to arrange for regular veterinary checkups for the animals. They have harvest the animals for their by products like milk, fur, meat etc. They have to arrange for buyers and get good prices for the goods

They have to make sure the business is making profits, and they have to maintain the accounts for the business. For the details, read the agribusiness manager job description given below.

Duties and responsibilities of an agribusiness manager

- They have to study and implement methods for plantation and harvesting
- They have to oversee all the processes like plantation and harvesting
- They have to arrange for pesticides in case the plant catch pests
- They have make sure the harvest is well stored and arrange for buyers
- If they package the harvest, the manger will oversee the packaging process too
- The manager has to manage the whole agriculture business, from production, to packaging, to sales, the manager is in charge of all
- They have to hire workers for they business and give them the adequate training as agriculture jobs require specialist training
- They also have to arrange for the workers employment in the off season

- The manager should have good contact with vendors that will sell them quality seeds at good prices
- The manger will look after all the accounts of the business and make sure that they are making profits
- They have to report to the owners of the business on the progress made

Qualifications Required

- A bachelor's degree in business administration or agriculture is pre-requisite
- Long experience working in the field of agriculture and related fields is desirable.

Agribusiness Manager Ratings

AAG's market leading research in the Agri Managed Investments Sector enables us to provide Manager Ratings. This data is derived from our alliance with Ernst & Young and their Corporate Governance Rating and AAG's own rating of Track record.

This is the first time that Manager Ratings have been compiled and represents another first for AAG and the agri MIS sector. Planners and investors can use this data to assist them in selecting Managers of agri tax effective products. Corporate Governance covers the following areas of review and rating: financial management, compliance, operational performance and governance. Track record covers the following areas of review and rating: management, past projects, markets, marketing and project financial performance.

How a business is managed and governed, and the results it has achieved over the time and fundamental pillars of Manager performance and an excellent guide to a quality Management team.

"The information available to planners and investors from AAG allows for a rational, commercial manager and hence investment selection", said Marcus Elgin, Managing Director of AAG. "We recommend however, that investors spread risk of management, commodity and geographic region of growing to create a blend of risk and return."

FUNCTIONAL CHARACTER OF AN AGRIBUSINESS

As a confirmed pragmatist, it is often difficult for me to abstract from the functional character of an agribusiness firm and, instead, view it as a social organization. To be sure, all businesses and the managerial control thereof, involve people. The very essence of any successful business rests on human interaction. But in our haste to analytically assess a business, to financially control its operations, and to quantitatively plan its future, we often forget to equate "business management" with the "leadership function" required of all social organizations.

One would normally expect that any institution-bound academician would spend much of his/her time in the simple act of reading. Unfortunately, this is one of many misconceptions which characterize the life of an academician. Not unlike agribusiness managers, academicians find that the normal "press of business" requires a vast diversity of actions, most of which prevent the individual from becoming deeply involved in the pursuit of leisurely reading. Unlike managers, however, academicians are occasionally afforded a brief respite from their routine press of business to read and pursue thoughts or ideas which would otherwise be judged inconsequential to their employer's day-to-day operational tasks.

As I pursue, and enjoy, just such a respite, I elect to read, and do so rather extensively. Not all my reading is focused on a specific subject, but much is directed towards my personal interests in the practice of business management. Acknowledging that agribusiness managers are rarely afforded even a brief respite from their daily tasks, I would like to share with you a condensation of observations expropriated from my current practice of leisurely reading.

We must be reminded of the fact that all organizations, social and business, are destined to failure in the absence of some form of leadership. Someone must be in charge. In its most vestigial form, this is even true in its natural form for the birds and the bees, the former with their pecking order, and the latter with their queens.

In the matter of human affairs, even those who would reject all forms of traditional leadership, find an undeniable need for leaders, themselves. I recently had an opportunity to debate with a Libertarian Party member of the Alaska state legislature.

His views were just short of those advocating complete anarchy in the affairs of state. Yet, ironically, it was this same party whose continued political influence was almost totally dependent upon a finely-tuned and well-disciplined party leadership structure.

The desire to equate leadership with management is an unavoidable one. Like cream, it would seem, leaders naturally rise to the surface where they fulfill managerial roles. Why is it then that managers, unlike cream, do not always represent the best part of the whole? To answer this question, we must recognize that the wizardry of popular leadership has, historically, contributed as much to evil as to good.

Adolph Hitler, for example, proved to be a charismatic leader of some renown, whose ability to stimulate a mass following for his own twisted visions surely propagated human suffering. In this noticeable case, when did leadership end and demagoguery begin? In a recent book titled Leadership, James MacGregor he wrote, "A leader and a tyrant are polar opposites." Quite clearly, Burns would distinguish between those who would lead and those who would abuse their position of influence.

Managing Agribusiness

This question was asked some months ago during a week-long agribusiness management training program. A team of five participants were asked to meet together as a team and produce a reasonable answer. It was expected that their answer would be lengthy and rather complex. In fact, they returned the following morning with a brief,but rather profound response. In their view, management required the organization of things, while leadership required the organization of people. In a book titled The Unconscious Conspiracy, author Warren Bennis wrote in 1976, "leading does not mean managing." Clearly, our team of management trainees and Mr. Bennis were hinting at a similar distinction.

It would seem that a manager who lacks the leadership quality would function mostly in accordance with previously established routines. Of course, such routines require little thought and even less imagination. They may, in fact, be inappropriate routines or they may be functionally obsolete. Yet managers lacking leadership skills will focus on things rather than people and cling to routine procedures as a defensive means for preventing needed changes.

A reliance on routines may not even be a conscious effort as we are all creatures of habit. Each of us seeks and sustains a sense of security through an adherence to established routines. When such an adherence is allowed to supersede a concern for people and the collective well-being of a business, then management ceases to be an analogue to leadership.

Leadership Role

Even those agribusiness managers who actively pursue a leadership role, find themselves inadvertently submerged in protectionist attempts to preserve the status quo. A noted management scientist named Henry Mintzberg once conducted an in-depth analysis of the working patterns of five top corporate executives. Mintzberg concluded that only rarely did these talented individuals have time to think about anything except the particular issue immediately before them.

Nearly half of the decisions made were rendered within a time period of ten minutes or less. In only ten percent of the cases did these individuals have over one hour to ponder and evaluate a specific matter.

In reality, the executive suite was but a frenzy of activity as a steady stream of correspondence and visits filled the entire working day. Under even the best of intentions to the contrary, these persons were forced to rely heavily on routines as the sole means for dealing with the press of business. If your daily work pattern is anything like that of the executives observed, it would seem as if you are running as fast as you can only to stay where you are. Under such conditions, how can anyone find the time and initiative to function as a leader? This question also arose in the course of the aforementioned management training program.

Perhaps, the answer lies in the manager's willingness and ability to ask himself if the use of routine has provided a subconscious excuse for avoiding the more difficult and demanding tasks.

LEADERSHIP PATTERN

In 1958, Robert Tannenbaum and Warren H. Schmidt published a research paper in the Harvard Business Review titled "How to Choose a Leadership Pattern." In this paper, the authors described a so-called "dominance scale" from which evolved alternative managerial styles. The anagement profession later accepted this dominance scale base for the classification of managerial styles which varied from Autocrat to Humanist. By 1973, however, these same authors felt compelled to produce a sequel to their earlier thoughts.

This sequel was prepared on the basis of perceived changes in social patterns that had taken place since 1958. The fifteen years that had passed since their earlier literary work had witnessed the rapid rise of concerns for civil rights, ecology, consumer protection and the quality of life in the workplace. In their opinion, such societal changes had placed upon managers a burden which previously had hardly existed, i.e., the heightened need for a sense of human sensitivity and flexibility in management.

What this suggests is that as a manager/leader in contemporary times, you are more likely to deal with employers with a higher self-image of their role in the business.

They are less willing to accept their position as a subordinate, more willing to question any source of authority, and more demanding of a position of influence in the business. In fact, a new breed of workers emerged from the 1970s period of social turmoil. While tempers have cooled somewhat and younger people have returned to more pragmatic interests, the modern day managers must recognize that now, more than ever before, leadership becomes a matter of eliciting support and cooperation from employees rather than commanding obedience. Recent studies show that employees are more concerned with a sense of personal autonomy, appreciation of their efforts, and an expanded opportunity to fulfill their individual potentials.

Essential Task of Management

If the manager fails to capitalize on the energies supporting these employee goals, those same energies are sure to become the business's loss. Management scholar Douglas McGregor argues that, "the essential task of management is to arrange organizational conditions and methods of operation so that employees can achieve their own goals best by directing their own efforts towards those objectives set by the organization." The managers of the "old school" might argue that such a permissive environment will only result in diminished productivity.

But allowing for the fact that all managers must, at times, play a dictatorial role, it would seem that employees would respond more positively to such moments of toughness, if it is encompassed within a past record of individual respect, consideration and fair play.

Conclusion

In conclusion, a rather simple rule emerges, i.e., the respectful treatment of others is likely to be reciprocated. And it is this reciprocation that transforms a manager into a leader. Lao-Tsi was a Chinese poet and philosopher who lived 2,500 years ago. He likely possessed little experience as a manager, yet his words of advice are still timely and relevant. He wrote, "Fail to honor people, and they will fail to honor you; but of a good leader, who talks little, when his work is done, his aim fulfilled, they will say: we did this ourselves."

Agribusiness Manager Job Requirements

There are a number of qualifications required to obtain a position as an agribusiness manager. They include:

Education: Most employers require agribusiness managers to have a bachelor's degree in agriculture, agriculture science or a related field. Some employers prefer to hire candidates with a master's degree in agriculture or a related field.

Many aspiring agribusiness managers choose to pursue a degree in agricultural economics. This degree provides students with a strong foundation in business and economics.

Training & Experience: Most agribusiness managers learn the specific skills and knowledge they need for their role while on the job. Training for this role often includes shadowing an experienced manager or working in a lower-level position before advancing to a managerial role. Training often includes learning about the company's products and services, the industry and the role's specific responsibilities.

Certifications & Licenses: Agribusiness managers are not typically required to have a certification or license to get hired, but some employers may prefer candidates who have one.

Agribusiness Manager Skills

Agribusiness managers need the following skills in order to be successful:

Leadership: Agribusiness managers often oversee a team of employees, so leadership skills are important for them to be successful in their careers. Leadership skills can include the ability to motivate employees, delegate tasks and inspire others to work hard.

Communication: Agribusiness managers communicate with a variety of stakeholders, including suppliers, customers, employees and investors. Effective communication is key to building trust and maintaining positive relationships. Agribusiness managers should be able to communicate clearly and concisely in both written and verbal forms.

Analysis: Agribusiness managers use analytical skills to assess the strengths and weaknesses of a business model, forecast market trends and evaluate the financial viability of potential business opportunities. They use analytical skills to develop business strategies and make informed decisions about the future of a company.

Problem-solving: Problem-solving is the ability to identify and address issues. As an agribusiness manager, you may be responsible for resolving issues that arise in the production of crops or livestock. You may also be responsible for finding solutions to challenges that arise in the supply chain, such as when a supplier is unable to fulfill an order.

Decision-making: Agribusiness managers make decisions about a variety of topics, including production, marketing and sales strategies. They also make decisions about the future of the company, such as whether to invest in new technology or machinery. Making effective decisions requires having a thorough understanding of the topic and the ability to weigh the pros and cons of each option.

Agribusiness Manager Work Environment

Agribusiness managers typically work in offices, but they may spend considerable time outdoors on farms or in other agricultural settings. They usually work regular hours, but they may be on call 24 hours a day to deal with emergencies. They may also travel to attend conferences or meet with clients. Some agribusiness managers work for agricultural consulting firms and travel frequently to meet with clients. The work of agribusiness managers can be stressful because they are responsible for the financial success of their businesses and the well-being of their employees.

3

THE ROLE OF MARKETING IN SUSTAINABLE AGRICULTURE

Agricultural marketing covers the services involved in moving an agricultural product from the farm to the consumer. Numerous interconnected activities are involved in doing this, such as planning production, growing and harvesting, grading, packing, transport, storage, agro- and food processing, distribution, advertising and sale. Some definitions would even include "the acts of buying supplies, renting equipment, (and) paying labor", arguing that marketing is everything a business does. Such activities cannot take place without the exchange of information and are often heavily dependent on the availability of suitable finance.

Marketing systems are dynamic; they are competitive and involve continuous change and improvement. Businesses that have lower costs, are more efficient, and can deliver quality products, are those that prosper. Those that have high costs, fail to adapt to changes in market demand and provide poorer quality are often forced out of business. Marketing has to be customer-oriented and has to provide the farmer, transporter, trader, processor, etc. with a profit. This requires those involved in marketing chains to understand buyer requirements, both in terms of product and business conditions.

In western countries considerable agricultural marketing support to farmers is often provided. In the USA, for example, the USDA operates the Agricultural marketing Service. Support to developing countries with agricultural marketing development is carried out by various donor organizations and there is a trend for countries to develop their own Agricultural marketing or Agribusiness units, often attached to ministries of agriculture.

Activities include market information development, marketing extension, training in marketing and infrastructure development. Since the 1990s, trends have seen the growing importance of supermarkets and a growing interest in contract farming, both of which impact significantly on the way in which marketing takes place.

THE FUNDAMENTALS OF MARKETING

New paradigms do not create new facts but instead provide new frameworks or filters for translating facts into knowledge. Likewise, the fundamentals of economic value are not new, but new opportunities and strategies for creating value through marketing arise from the sustainability paradigm. Effective marketing begins with a basic understanding of the fundamental realities of markets, regardless of one's paradigms.

Marketing, in agriculture, includes all the various activities involved in the transformation of commodities sold by farmers into food and fiber products purchased by consumers. The most obvious aspect of this transformation is a change in physical appearance or form. Form changing activities for agricultural commodities range from washing and grading apples to processing wheat into Wheaties. Another important marketing function is transportation. Agricultural commodities must somehow get from the farms where they are grown to the retail outlets where they are bought, in some cases moving across a country or half-way around the world.

Time is another important aspect of marketing. Many agricultural commodities must be harvested at a specific time, but can be stored for later use and in some cases be consumed year-round. Finally, in a specialized economy, most consumers are not producers. Marketing involves the transfer of ownership or possession from those individuals who produce, ultimately, to those individuals who consume the food or utilize the fiber.

Market transformations change the value of commodities by changing their form, place, time, and ownership as they move through the marketing system. Each of these functions creates value but also adds an associated cost. Profits result whenever the value added by marketing functions such as processing, transportation, storage, or brokerage, is greater than the costs of performing those functions. These basic principles of marketing may seem simplistic; however, they are the fundamental concepts upon which vertical expansion of farming into marketing must be built.

Marketing, to most farmers, means commodity marketing. They produce commodities such as corn, wheat, hogs, or cattle. One farmer's No. 2 grade yellow corn is pretty much like any other farmer's No. 2 yellow corn. One cattle feeder's 1100 lb. USDA choice steers are a lot like steers of a similar weight and grade from any other feed lot. Thus, the commodities one farmer has to offer for sale are freely interchangeable with commodities offered for sale by many other farmers, oftentimes including farmers on another continent. Commodity markets tend to be highly competitive because there are many buyers and sellers of the same basic commodity. Price differences among different market locations rarely exceed transportation costs and price changes over time tend to reflect seasonal or cyclical cost differences.

Commodity marketing decisions are primarily limited to decisions of timing. Farmers can forward price their commodities through private contracts or futures markets, attempting to get a price higher than market price at time of delivery. Or they can store commodities for later sale, hoping that market prices will rise more than their costs of storage. In either case, farmers are matching wits with speculators who make their living buying, pricing, storing, or selling commodities. Most conventional farmers are not particularly good speculators. They make a living by keeping their costs competitive and, thus, being able to stay in business at competitive market prices.

Product marketing is different from commodity marketing. Commodities are alike, but products are different. In marketing jargon, products have distinct quality characteristics and, thus, distinct market values. These differences may be tangible in nature (as in nutrient values of foods) or intangible (as in consumer acceptance created by brand advertising). Differentiation creates a more or less unique market for a product, taking it out of direct competition with other products. The greater the differentiation, the greater the potential for profits. Products that have few good substitutes may command a substantial price premium over less acceptable alternatives. However, consumers will not pay much more for a product that has many good substitutes than they will pay for the substitutes.

Products may be differentiated by anything that affects value. Processing, for example, changes the form of raw commodities. Some processing activities, cattle slaughter for example, result in a different form of generic commodity, such as choice beef. In other cases, however, processors are able to transform commodities into distinct products such wines from grapes. A supermarket may stock wines from a number of different wineries because the wines are perceived to be different by a sufficient number of wine drinkers to create profitable markets. Processing is only one means of changing the form of a commodity. Fruits sorted for uniformity of size and color, for example, may sell for a premium over the same fruit sold unsorted.

Products may also be differentiated with respect to time. In mid-winter, local greenhouse tomatoes in the Midwest may command a substantial premium over tomatoes from California or Mexico. However, tomatoes from those same greenhouses may have no advantage over local vine-ripe tomatoes in mid-summer. Winter price premiums reflect the lack of good substitutes at that particular time. Vine-ripe tomatoes out-of-season are a differentiated product, but in-season tomatoes are a commodity. An ability to provide products on time, continuously over time, may also differentiate one farmer's products from the others.

Location is another factor which differentiates the value of products. Farmers near population centers have a distinct advantage in most direct marketing strategies. Pick-your-own fruit, vegetable, or berry farms, for example, must be located within reasonable driving distance of a significant population center. Farmer's markets are also logical market outlets for producers of fresh produce from a fairly limited geographic area. Product handling and transportation technologies, however, have reduced the significance of location, either as an advantage or as an obstacle in marketing.

Market advantages associated with individual ownership or possession are perhaps less obvious and less well understood than those associated with form, time, and place. However, values associated with individuality may be far more important than any other in developing sustainable agricultural systems. Different individuals, or groups of individuals, value the same products differently. In other words, the same form or physical quality of product may be valued differently by two different groups of people at any given time and place.

Thus, individual farmers can command higher market prices simply by offering their products to the individuals or groups who value them most. Products that are carefully tailored or targeted to meet the specific needs of narrowly segmented markets may command a significant price premium over mass-produced commodities that meet the same generic need. Matching products to the tastes and preferences of specific consumer groups is the essence of successful niche marketing.

Scope for Agribusiness in India

India is endowed with varied ago-climate, which facilitates production of temperate, sub-tropical and tropical agricultural commodities.

There is growing demand for agricultural inputs like feed and fodder, inorganic fertilizers, bio-fertilizers.

Biotechnology applications in agriculture have vast scope in production of seed, bio-control agents, industrial harnessing of microbes for bakery products.

Export can be harnessed as a source of economic growth. As a signatory of World Trade Organization, India has vast potential to improve it present position in the World trade of agricultural commodities both raw and processed form. The products line include cereals, pulses, oilseeds and oils, oil meal, spices and condiments, fruits and vegetables, flowers, medicinal plants and essential oils, agricultural advisory services, agricultural tools and implements, meat, milk and milk products, fish and fish products, ornamental fish, forest by products etc.

At present processing is done at primary level only and the rising standard of living expands opportunities for secondary and tertiary processing of agricultural commodities.

The vast coastal line and internal water courses provides enormous opportunity for production of marine and inland fish and ornamental fish culture gaining popularity with increase in aesthetic value among the citizens of India..

The livestock wealth gives enormous scope for production of meat, milk and milk products, poultry products etc

The forest resources can be utilized for production of by products of forestry.

Beekeeping and apiary can be taken up on large scale in India.

Mushroom production for domestic consumption and export can be enhanced with improvement in the state of art of their production.

Organic farming has highest potential in India as the pesticide and inorganic fertilizer application are less in India compared to industrial nations of the world. The farmers can be encouraged and educated to switch over for organic farming.

There is wide scope for production and promotion of bio-pesticides and bio-control agents for protection of crops.

Seeds, hybrid and genetically modified crops, have the highest potential in India in the future, since the productivity of high yielding varieties have reached a plateau.

Micro-irrigation systems and labor saving farm equipments have good potential for the years to come due to declining groundwater level and labor scarcity for agricultural operations like weeding, transplanting and harvesting.

Production of vegetables and flowers under green house conditions can be taken up to harness the export market.

Trained human resources in agriculture and allied sciences will take on agricultural extension system due to dwindling resources of state finance and down sizing the present government agricultural extension staff as consulting services.

The enhanced agricultural production throws open opportunities for employment in marketing, transport, cold storage and warehousing facilities, credit, insurance and logistic support services.

The Industrialization of Agricultural Markets

Questions of sustainability arise primarily in relation to the consequence of industrialization. An understanding of this linkage between sustainability and industrialization is critical in planning for a sustainable future, regardless of whether the specific subject is marketing, production, or public policy. U.S. agriculture was industrialized to support industrialization of the economy in general. At the turn of the century, America was an agrarian society. Most of our total resources were spent on the necessary tasks of feeding and clothing ourselves. People had to be freed from producing food and fiber to provide workers for the factories and offices of a growing industrial society. The costs of food and fiber had to decline if consumers were to have discretionary income to buy the things that the factories and offices would produce. Agricultural industrialization was a logical means for meeting both objectives.

U.S. agriculture was mechanized, specialized, routinized, and otherwise industrialized making it possible for fewer people to feed more people better for less money. Industrialization began in earnest when tractors replaced horses as sources of power on U.S. farms. The agricultural sector has been among the last to become fully industrialized, but the driving force of change has long been to make farms perform as factories without roofs with fields and feed lots operating as much as possible like factory assembly lines.

The industrial approach to marketing is mass marketing to support mass production. Factories must turn out large quantities of identical, or quite similar, units to achieve the economies of scale associated with assembly-line production. Automobile manufacturing is a prime example of industrial production, and the black Model T Ford is still the "classic" industrial product. Today, automobiles come in a variety of sizes, shapes, colors, and extras, but they are still, with few exceptions, basic commodities produced for mass markets. Millions of dollars are spent on advertising to ensure that masses of consumers will be willing to drive automobiles that can be mass produced by industrial methods. Farms, like factories, produce large quantities of basic commodities -- corn, wheat, soybeans, hogs, cattle -- for mass markets. Agriculture is promoted among consumers as a supplier of low-cost, basic food stuffs to feed the hungry masses.

Industrial strategies of the past century were highly successful in reducing the claim of food and fiber production on the nation's human resources. For example, the 1895 Yearbook of Agriculture indicated that 42 percent of people in the U.S. were employed on farms in 1890 (USDA, 1895). This compares to less than 2 percent of the total U.S. population living on farms a century later. In addition, those living on farms today earn more than half of their income from non-farm sources (USDA, 1990). U.S. consumers now spend just over 11 percent of their income on food, and farmers get only about 22 cents out of each dollar spent for food (Dunham, 1993). About half of the farmer's share goes to pay for purchased inputs, leaving the farmer little more than a dime out of each dollar spent for food.

Paradigms, such as industrialization, become dominant because they are found to be capable of solving problems that previous paradigms could not solve (Baker, 1992). The industrial era, in general, was fostered by a host of interrelated and complex developments, including access to large supplies of affordable fossil fuels. The industrialization of agriculture was made possible by these same developments, but was driven by the nation's necessity for fewer farmers to feed more people at a lower cost. The industrial paradigm succeeded. Food is cheap in the United States, and the farmer's share of food costs is even cheaper. Farmers and other rural people have been freed from the subsistence living that characterized earlier times. The problems agriculture was industrialized to solve have been solved. If farmers took nothing for their part in food production, average cost of food would only be about ten percent less. And, there aren't many farmers left to move off the farms, if there was anywhere else for them to go.

The initial logic for industrializing agriculture is no longer valid. In addition, the industrialization of agriculture has generated a whole new set of growing economic, environmental, and social concerns that may already outweigh its declining benefits. A growing number of people are looking to a fundamentally different paradigm as they search for answers to the economic, ecologic, and social questions arising from industrial methods of farming. These people need a new paradigm for the creation of economic value, a paradigm that includes marketing as well as production.

AGRICULTURAL MARKETING DEVELOPMENT

A typical market in AfricaWell-functioning marketing systems necessitates a strong private sector backed up by appropriate policy and legislative frameworks and effective government support services. Such services can include provision of market infrastructure, supply of market information as done by USDA, for example, and agricultural extension services able to advise farmers on marketing. Training in marketing at all levels is also needed.

One of many problems faced in agricultural marketing in developing countries is the latent hostility to the private sector and the lack of understanding of the role of the intermediary. For this reason "middleman" has become very much a pejorative word.

Agricultural Advisory Services and The Market

Promoting market orientation in agricultural advisory services aims to provide for the sustainable enhancement of the capabilities of the rural poor to enable them to benefit from agricultural markets and help them to adapt to factors which impact upon these.

As a study by the Overseas Development Institute demonstrates, a value chain approach to advisory services indicates that the range of clients serviced should go beyond farmers to include input providers, producers, producer organisations and processors and traders.

Market Infrastructure

Efficient marketing infrastructure such as wholesale, retail and assembly markets and storage facilities is essential for cost-effective marketing, to minimise post-harvest losses and to reduce health risks. Markets play an important role in rural development, income generation, food security, developing rural-market linkages and gender issues.

Planners need to be aware of how to design markets that meet a community's social and economic needs and how to choose a suitable site for a new market. In many cases-sites are chosen that are inappropriate and result in under-use or even no use of the infrastructure constructed.

It is also not sufficient just to build a market: attention needs to be paid to how that market will be managed, operated and maintained. In most cases, where market improvements were only aimed at infrastructure upgrading and did not guarantee maintenance and management, most failed within a few years.

Rural assembly markets are located in production areas and primarily serve as places where farmers can meet with traders to sell their products. These may be occasional or weekly markets, such as haat bazaars in India and Nepal, or permanent. Terminal wholesale markets are located in major metropolitan areas, where produce is finally channelled to consumers through trade between wholesalers and retailers, caterers, etc. The characteristics of wholesale markets have changed considerably as retailing changes in response to urban growth, the increasing role of supermarkets and increased consumer spending capacity. These changes require responses in the way in which traditional wholesale markets are organized and managed.

Retail Marketing

Retail marketing systems in western countries have broadly evolved from traditional street markets through to the modern hypermarket or out-of-town shopping centre. In developing countries, there remains considerable scope to improve agricultural marketing by constructing new retail markets, despite the growth of supermarkets, although municipalities often view markets as sources of revenue rather than infrastructure requiring development. Effective regulation of markets is essential. Inside the market, both hygiene rules and revenue collection activities have to be enforced.

Of equal importance, however, is the maintenance of order outside the market. Licensed traders in a market will not be willing to cooperate in raising standards if they face competition from unlicensed operators outside who do not pay any of the costs involved in providing a proper service.

Market Information

Efficient market information can be shown to have positive benefits for farmers and traders. Up-to-date information on prices and other market factors enables farmers to negotiate with traders and also facilitates spatial distribution of products from rural areas to towns and between markets. Most governments in developing countries have tried to provide market information services to farmers, but these have tended to experience problems of sustainability. Moreover, even when they function, the service provided is often insufficient to allow commercial decisions to be made because of time lags between data collection and dissemination. Modern communications technologies open up the possibility for market information services to improve information delivery through SMS on cell phones and the rapid growth of FM radio stations in many developing countries offers the possibility of more localised information services. In the longer run, the internet may become an effective way of delivering information to farmers. However, problems associated with the cost and accuracy of data collection still remain to be addressed. Even when they have access to market information, farmers often require assistance in interpreting that information. For example, the market price quoted on the radio may refer to a wholesale selling price and farmers may have difficulty in translating this into a realistic price at their local assembly market. Various attempts have been made in developing countries to introduce commercial market information services but these have largely been targeted at traders, commercial farmers or exporters.

It is not easy to see how small, poor farmers can generate sufficient income for a commercial service to be profitable although in India a new service introduced by Thompson Reuters was reportedly used by over 100,000 farmers in its first year of operation. Esoko in West Africa attempts to subsidize the cost of such services to farmers by charging access to a more advanced feature set of mobile-based tools to businesses.

The Business of Paradigms

Joel Barker, in his book Paradigms, defines a paradigm as a set of rules that do two things: (1) establishes or defines boundaries and (2) sets standards for success and behavior within the boundaries. He uses the game of tennis as an analogy to illustrate these concepts. Tennis courts are standard in size and out-of-bounds are clearly marked. The ball must hit within these bounds to "stay in play." The ball must be struck with a tennis racket, not a baseball bat or anything else, and the ball is allowed to bounce only once before it is returned over the net.

Paradigms may be simple, as in the case of games, or extremely complex, as in the case of a model for economic development. However, the industrial model has some clearly defined boundaries. The natural environment, including the natural resource base, is considered to be "external," or out of bounds, by industrial managers. Society likewise is considered to be an "external" factor which constrains or sets bounds on what industrial firms can do. Success for an industrial firm is measured in terms of profits and growth. Within the limits allowed by nature and society, industrial firms may take a wide range of actions to maximize short run profits and long run growth. Almost anything that is possible and legal is encouraged if it leads to profits and growth.

Profits and growth are seen as a natural consequence of reducing costs and expanding market share. Industrial firms must be cognizant of consumer preferences, but they search for large groups of consumers who are willing, or can be persuaded, to buy the same basic item. Industrial firms need new mass markets to keep their production cost low and their profits up as they continually strive to expand the scale of their operation.

The new paradigm emerging under the sustainable agriculture umbrella may not be fully understood for some time to come. However, this post-industrial approach to farming is fundamentally different from the industrial paradigm in several ways. The new paradigm for agriculture clearly considers ecologic and social impacts to be "within" rather than "outside" of its boundaries. Ecological soundness and social responsibility are seen as positive goals rather than negative constraints. The new constraints or boundaries are the laws of nature, including human nature.

The new paradigm considers economic, ecologic, and social dimensions of sustainability to be inseparable. Fields, farms, communities, and societies are considered to be wholes that are made up of smaller wholes and make up still larger wholes. Thus, the approach to farm decision-making and management must be "holistic." The challenge is to comprehend the complexities of wholes rather than attempt to reduce wholes to more simple and easily understood elements. Success in the new paradigm is measured against the goal of sustainability rather than profits and growth.

Progress is seen as a consequence of serving the needs and wants of people by means that are ecologically sound and economically viable. Site-specificity is generally accepted as a prerequisite for sustainable farming. Farming systems must be matched with their natural resource base. Sustainable farming systems also must be individualistic. Management and labor requirements must match the talents, abilities, and aspirations of individual farmers. It seems logical that sustainable systems also need to be market specific. Unique sets of production resources, both natural and human, need to be matched with unique groups of consumers -- unique markets -- if systems are to be sustainable.

The new sustainable model implies greater reliance on human resources, in terms of the quality and quantity of labor and management, and relatively less reliance on land and capital. Industrial systems are management extensive, requiring fewer managers per acre or per dollar of capital investment, and little if any management input from hired workers. Sustainability systems, on the other hand, are more management extensive, employing more managers per acre or per dollar of capital investment, and more management participation from hired workers. Thus, successful farming with the new sustainable paradigm would require more farm operators, more productive farm workers, and more farm families than does an industrial system of farming.

Smaller, more diversified farms would become the norm under the sustainable paradigm. However, smaller, diversified farms will be commercially competitive with larger specialized farms only if human resources can be substituted economically for other on-farm resources and commercial inputs. By implication, farmers who succeed with the new paradigm must be more productive as "people." They cannot expect to earn a larger return for their management and labor, thereby reducing their reliance on land and capital, unless they possess unique skills and abilities. Production skills will be critical to overall ecological and economic viability. However, marketing skills may be even more important than production skills in maintaining the economic viability of smaller, diversified farms in the future. The successful employment of people in production and marketing will be critical in maintaining a socially responsible agriculture.

Much of the past emphasis in sustainable agriculture has been on reducing reliance on purchases inputs. However, as indicated previously, input purchases account for about a dime of each dollar consumers spend on food while marketing services account for eighty cents. It seems logical that farmers have more to gain, at least economically, by attempting to capture part of the eighty cents spent for marketing services rather than focus on the dime spent for inputs. However, even greater opportunities may be found in strategies which enhance the final value of products rather than reduce costs of inputs or marketing services. Smaller farmers may face even greater difficulties competing with industrial firms as providers of marketing services than they have competing as producers. The key to success in marketing products from smaller, diversified farms may be the same as for overall sustainability -- a new paradigm.

AGRICULTURAL MARKETING IN DEVELOPING COUNTRIES

Economic reforms have had sweeping impacts on agricultural markets in developing countries. In general, state intervention has been reduced, notably with respect to:

- The abolition or sharp curtailing of parastatal marketing boards;
- Depreciation of formerly over-valued currencies rendering developing country exports more competitive and imports more expensive;
- A reduced public role in agricultural services, especially in subsidized credit, input and extension networks; and
- A shift away from pan-territorial and pan-seasonal crop pricing strategies and pre-announced prices.

Reviewing agricultural markets research in sub-Saharan Africa and Asia, Jones (1996) concludes:

"In newly liberalized [food] markets in eastern and southern Africa . . . barriers of entry to trade are low, but the marketing system has little capacity to channel credit or spread risk. There are strong theoretical reasons for expecting the impact of, and response to reforms to vary between different classes of producers. The absence of key markets, risk aversion, high transaction costs and the dual role of agricultural households as producers and consumers are critical features. The marketing system depends on both physical and institutional infrastructure . . . Collective action by market participants may address this but it may also lead to collusion over prices . . . Evidence from South Asia shows that food markets exhibit social barriers to entry, massive asset polarization, debt relationships between large and small traders and traders and farmers, diverse institutional and contractual arrangements, and collusive behaviour, enforced in part by manipulation of the state regulatory system."

This conclusion gives some clue to the reasons why NGOs and CBOs intervene in agricultural markets. When extension agents, researchers and development organizations working in rural areas ask farmers to prioritize their problems, agricultural marketing is repeatedly raised as one of the most important problems faced. It may arise in the context of the promotion of new crops or productivity-enhancing technology, or it may be felt particularly acutely in remoter areas poorly served by commercial traders, where parastatals no longer operate. NGO marketing interventions typically aim to fill critical gaps in the marketing system or address the power imbalances.

Nowhere are those marketing problems felt more acutely than in the areas for which it is most difficult to identify sustainable strategies to improve market access. Farmers in remote areas (either remote because of physical distance from markets or because of poor roads) are almost always poorly served by agricultural traders and are often obliged to accept seemingly unattractive prices for their produce.

Distance from markets rules out the production of higher value more perishable crops, and reduces the linkages between these producers and other more specialized markets. By the same token, CBOs and NGOs seeking to promote alternative strategies for these disadvantaged communities face high costs and tangible obstacles that make their task particularly difficult.

Poor access to markets is mirrored by poor access to all kinds of rural services. The poverty that results makes such communities particularly risk-averse. Where rainfall is uncertain, the situation is even worse, whilst the relative absence of trade does nothing to relieve the covariance in production.

These are the challenging circumstances that make an examination of marketing interventions worthwhile. There is wide-ranging experience amongst the development NGO community. Some of these initiatives have taken-off and developed into self-sustaining activities, whilst others, although not conceived as such, have effectively become subsidy dependent welfare programmes. This review identifies best practice and the conditions required for such programmes to work.

NGO AND CBO INVOLVEMENT IN AGRICULTURAL MARKETING

Different Types of Organization

1. Indigenous NGOs who have become relatively large, well-organized and able to attract significant funds from international donors and Northern NGOs; sometimes these NGOs were originally created or strengthened by Northern NGOs.
2. Northern NGOs with offices in developing countries, usually obtaining funds from donors (including private individuals); this group is quite broad since it encompasses large NGOs such as Oxfam or CARE, with activities in many countries, as well as small NGOs whose activities may be quite focused on a few countries and issues.
3. CBOs, membership organizations serving particular interest groups usually in rural communities, whose focus may be broad or quite narrow; these organizations may be formally structured or quite informal; farmers associations, credit groups, and joint marketing societies could all be considered CBOs in the context of this review.
4. Indigenous NGOs that are small, usually focused on a particular geographical area or issue, that obtain small amounts of funding from donors or government, but who struggle to grow or stay afloat.

Many countries have laws governing the registration of different types of organizations that may confer a certain tax status or legal standing. Some developing countries have umbrella associations for NGOs. In any particular country, it is useful to find out whether an umbrella organization exists, and if so, the types of CBOs and NGOs that tend to be officially registered – recognizing the potential divergence between official requirements and practice.

The Evolving Role of Ngos and Cbos in Development Assistance

In the last two decades, NGOs have become progressively more involved in development assistance, at every level. The shift from a relief and welfare focus has come about partly in an attempt to address the underlying causes of some of those man-made disasters or to limit the negative consequences of the natural disasters at which they assisted. It has been helped by the increased funding they found they were able to attract.

Many Northern NGOs now have policy and research departments, and are a legitimate channel for large amounts of donor funding. At the same time, the role of the state has been redrawn, and in developed and developing countries, there is now a much greater focus on civil society as a way to improve democratic processes and bring about greater accountability in government.

Governments are also seeking ways to be smaller and to sub-contract functions where feasible. Furthermore, funding developing country organizations to carry out development work is considered a way to build indigenous capacity.

NGOs working in developing countries have benefited from this trend – either because they are considered part of civil society or because they work closely with many civil society organizations, including CBOs. As Stocker and Barbor-Might (1999: 3) state:

"From the point of view of the donors, civil society was the 'place'where something could be done and, often enough, NGOs were the intermediary institutions or midwives of such remedial programmes relating to structural adjustment, spanning the gap between donors and CSOs. The funding channels varied, sometimes being directed through Northern NGOs (which might provide 'aid' directly or channel it to one or more partner Southern NGOs or CSOs) and sometimes going as direct funding to Southern NGOs and in some instances even to CSOs. When governments were irredeemably corrupt or oppressive (as, for instance, in Haiti during the Duvalier regime), these programmes seemed to offer virtually the only hope of channelling assistance to the people who most needed it.

This growth in the funding, remit, competencies and responsibilities of NGOs has meant that they have been closely involved in if not the instigators of much of the experimentation with practical solutions to pressing problems in rural areas. This is the context in which NGO experiences with agricultural marketing interventions provide a valid and rich focus for this review.

Role of NGO in Agricultural Marketing

A large number of NGOs and CBOs become involved in agricultural marketing activities, but this is rarely their core business. Although many NGOs share similar altruistic goals, their approaches vary enormously. This is particularly evident in the extent to which they embrace and harness commercial activities to promote broader objectives, or reject this as a legitimate means by which to achieve social objectives.

Moreover, amongst those NGOs prepared to use commercial activities as a means to an end, there can be considerable variability in the role these activities are accorded within the development strategy and the competence with which they are planned and undertaken.

Organizations That Are Primarily Welfare-oriented

There are some notable exceptions amongst some of the international NGOs who have become very experienced in agricultural enterprise, agro-processing and marketing. These include, for instance, Techno Serve, the Intermediate Technology Development Group, Enterprise Works Worldwide and the Cooperative League of the USA.

Many NGOs start with welfare (or social or altruistic) objectives, in areas such as education, health, water, infrastructure and agriculture and gradually shift towards a longer-term development focus. With this shift, small business and income-generation activities take on a greater role. Gibson (1993) describes this gradual transformation in terms of a continuum of different actions and attitudes.

Often NGOs and CBOs deliberately work in remote and disadvantaged communities and target the poorest households or individuals. These conditions, in combination with a general relief and welfare orientation, influence the strategies they adopt to achieve their objectives.

For instance, direct or indirect subsidies may be used to improve access to markets (e.g. through the provision of transport, credit or inputs). An example of a direct subsidy is free or below cost use of transport (calculated on the basis of costs of fuel and driver and perhaps some portion of the vehicle costs). An indirect subsidy might involve charging a commercial or break-even rate on the vehicle hire but taking no account of the staff costs of implementing and managing the scheme. Whilst few people would suggest that the subsidy could continue indefinitely, there may be little consideration of how these activities can eventually be shifted to a more sustainable basis.

The result is often that the programme attracts participation because of the subsidies, and once it ends there is little enduring impact.

Yet in the short-run these types of activities are attractive to NGOs because they have fairly immediate and visible if not enduring impacts and can with varying degrees of success be targeted to particularly disadvantaged groups such as the poorest households, women, refugees, the handicapped or other marginalized social groups. An approach that seeks to use commercial channels may take much longer to develop and may place the intended target group at a disadvantage relative to other members of the community.

Business-like NGOs and CBOs

Organizations which themselves resemble small businesses – in terms of their people, culture, systems, structure and behaviour – are most likely to be successful in encouraging the growth of small businesses.

In recent years, private sector development has increasingly been seen as a viable and important approach to sustainable development. Thus, many governments, NGOs and CBOs have focused on the promotion of marketing and small-scale enterprise to encourage greater participation in the commercial sector, as a route to higher incomes, employment generation and growth.

Small enterprise development work, which grew considerably in the 1980s, has contributed to a realization that it is possible to make much greater use of market mechanisms in pursuit of development objectives. This has been shown particularly through the success of micro-credit initiatives, where even very poor individuals are able to repay not only loans but also interest which sometimes covers the costs of providing credit.

Social & Commercial Objectives

Also, social objectives and commercial objectives are not mutually exclusive and many NGOs and CBOs pursue both. The fair-trade movement is a good example of this. Fair-trade organizations use commercial methods to generate social development benefits through improved terms of trade. The important thing is to balance potential marketing success with the social benefit needs of the beneficiaries.

Furthermore, selective use of subsidies can still lead to sustainable and successful marketing initiatives, depending on the circumstances, as demonstrated by the CARE Egypt Agricultural Reform Programme. The programme provides information services to smallholder farmers and facilitates linkages to help increase farmer income. The service is highly subsidized but has proven successful for a number of reasons:

- It helps to link farmers to sources of information outside the programme, thereby fostering the long-term sustainability of relationships and networks; and
- Farmers contribute financially, i.e. they pay fees for the services; . the demand for services is farmer-driven and project staff work with farmers to identify production and marketing opportunities.

INCOME-GENERATING PROGRAMME

The marketing role that NGOs and CBOs take on lies somewhere along a continuum between being directly responsible for marketing activities to facilitating beneficiaries/clients to market for themselves.

Direct Marketing Role

The term 'income-generating programme' (IGP) is used to describe a variety of programmes. These range from enterprises owned and managed by the beneficiaries to enterprises owned and managed by the organization which employs the beneficiaries. A number of NGOs/CBOs have established this latter type of small business to generate income to finance their other programmes and reduce donordependence.

CBOs and NGOs can also be more directly responsible for marketing activities. One way of doing this is through outgrower schemes (sometimes referred to as contract farming or satellite production). Such schemes involve smallholder producers providing agricultural raw materials to trading or processing businesses. Often growers work as a group, characterized by Stringfellow et al. (1997) as linkage-dependent groups. Generally the marketing arrangements are pre-determined: prices or a pricing formula are agreed.

They help markets function to the benefit of both producers and the companies or organizations involved. This type of relationship is beneficial for farmers because they have a secure market for their produce at a predetermined price and the buyer benefits from having a guaranteed source of raw materials and lower transaction costs, which reduces his/her risk and costs.

Within the fair-trade arena, the role of, and the marketing channels used by NGOs and CBOs (or alternative trading organizations – ATOs), also varies. Some organizations such as Oxfam Trading and Traidcraft take a direct marketing role by acting as wholesalers, with the producers acting as subcontractors producing to order. An advantage of this type of arrangement for producers is that they are guaranteed a volume of sales, thereby minimizing their risk.

A disadvantage of this, and of out-grower schemes, is that the producers can be dependent on the trader, and may not have access to alternative buyers or markets if for any reason the trader is no longer able to market their produce.

Facilitative Role

Other NGOs and CBOs play a more facilitative role. They assist individuals, groups and communities to market for themselves. This includes both improving access to, and benefits generated from, existing products and existing markets as well as creating new products and new markets through technology development and processing.

There are a variety of ways in which organizations facilitate marketing, including: strengthening the capacity of individuals, groups or communities (through group strengthening and training); developing linkages to traders and other stakeholders in the marketing chain (e.g. input suppliers, credit sources and transport agents); and linking farmers to relevant market information.

This type of facilitative role is beneficial for a number of reasons: being less interventionist, it is likely to generate more sustainable marketing activities and linkages; it is likely to be achieved at lower cost than if the NGO was more directly responsible for marketing activities; and, therefore, it facilitates reaching a wider audience.

Types of Marketing Intervention

There are many different ways in which NGOs or CBOs may intervene to improve access to agricultural markets. In this study, interventions are discussed in eight non-exclusive categories that describe aspects of the intervention strategy:

- Marketing linkages
- Credit programmes
- Marketing information
- Holistic approaches
- Intended beneficiaries
- Skills and training
- Access to agricultural inputs
- Agro-processing technologies

Common Themes

MARKETING EXPERIENCES

In the previous study, NGO and CBO experiences with agricultural marketing interventions were reviewed, and although the eight topics covered were somewhat different, a number of common themes emerge from these experiences.

NGOs and CBOs increasingly see business development and promotion of viable commercial activities not just as a legitimate way to achieve broader social objectives, but also as a means by which the benefits generated may reach a wider audience and be sustained.

This commercial focus cannot be pursued effectively unless the NGO/CBO is also 'business-like' in its approach, employing people with the appropriate skills, adequately resourcing these activities, and according them status in their programmes commensurate with their potential to generate significant sustainable benefits.

It is important to build on existing marketing channels, rather than attempt to circumvent them or establish new ones. Existing channels will outlive the project and a sustainable strategy is more likely to be one that enables rural communities to achieve more from what is already there, rather than to try to replace it.

By the same token, more facilitative and less interventionist strategies are likely to be more sustainable.

The discussion of groups or individuals as a target audience was inconclusive. Groups can offer a cost-effective vehicle for service Any particular marketing intervention may comprise elements from several categories like inputs and training, or technology, training and finance.

MARKETING MANAGEMENT

The world is facing a growing gap between food supplies in transitional countries and the rest of the world. While some countries face surplus food production - others face famine. The world agricultural system has become a single market and important developments are reshaping the competitive field for all players. Some organisations are working out how and where they fit, while others are left behind.

The Agribusiness, Post Harvesting and Marketing Management training programme aims to assess the global agricultural system from the perspective of genomics, governments, farms, supermarkets and consumers.

One of the biggest problems that transitional countries face is how to market their hard grown agricultural products, both inland and for export. Professional agricultural marketing is regarded as one of the ways to overcome seasonal agricultural surpluses, shortages of food supplies and also as a means of generating more income.

The combination of these aspects in one training programme is derived from the ever growing need for all small farmers, suppliers to marketing firms and processing companies to better interact with each other to the mutual benefit of everyone concerned.

Successful marketing of agricultural products is dependent on the creation of favourable circumstances as well as the provision of resources and services. Effective marketing channels have to include infrastructure and facilities such as storage, handling, transporting, processing, packaging and retailing services. Successful marketing must also include product location, timing, product, quantity, quality, prices and all other information required by producers and consumers to make beneficial decisions.

Investment must also be made in research and development of product variety, post-harvest and processing, farm mechanisation as well as in food quality control. Research and development will be discussed extensively in this programme as one of the most influential components for the business success of each farmer.

Marketing in the Niches

A niche is analogous to a corner, nook, crack, or cranny. A niche represents something that is seen as being different, uncommon, or exceptional. A niche is small only because it is defined in comparison to something larger. A crack in a sidewalk, a niche in one thing, may be far bigger than a grain of sand, the whole of something else. The essence of a niche is its uniqueness relative to its physical surroundings or conceptual context. Thus, a niche market may be defined as a market that is different from the predominant market for the same basic commodity or generic class of products.

Market niches may be identified in terms of form, time, place, or individuality. For example, the market for blue corn may be considered a niche market. The market for local-fresh produce in season may be a niche market. Or Saint Louis, Mo might be a niche market for certain ethnic foods. However, market niches are associated, most fundamentally, with specific individuals or groups of consumers.

The people who prefer blue corn chips make up a niche in the corn chip market only because most people prefer chips from white or yellow corn. The people willingly pay more for local-fresh produce in season represent a niche in the fresh produce market, not the particular time of year. And the ethnic communities of people in Saint Louis, not their geographic location, make up niches in the mass consumer market of the Saint Louis region. Niche marketing is about identifying and serving individuals or identifiable groups who have unique tastes and preferences for products that are different from those of the larger populations of which they are part. Niches are often "described" using differences in form, place, or time of delivery. But, market niches are "defined" by differences in tastes and preferences among people.

Industrialization creates market niches. This assertion is based on the premise that people differ in their tastes and preferences. If this is true, we still have different underlying tastes and preferences, even though we have been bribed and persuaded to buy the same basic things to accommodate the needs of an industrial economy. We have been bribed by the lower prices and persuaded by advertising and promotion.

By accepting pretty much the same things as other people, not only because we can be "in style," but we can be "in style" at a lower price. Enough people were willing to accept a black Model T Ford to allow ordinary people to ride in style. But, that did not mean that everyone "preferred" a black Model T. Those who were not totally satisfied represented market niches for other car makers. Some niches grew into mass markets, but others did not. For many people, managing a riding stable may still be more rewarding, personally and financially, than managing an automobile dealership.

The industrialization of agriculture quite likely has created a multitude of niches in markets that are largely undiscovered. Food consumers have been bribed and persuaded into buying mass produced foods through lower prices, advertising, and other forms of promotion. Mass manufacturing and distribution systems employed in food marketing have limited consumer choices to those items that can compete for scarce space in shelves of modern supermarkets. For example, consumers have no choice other than USDA choice grade beef, or a single similar quality store brand, in most supermarket meat cases. Fruits and vegetables, while abundant in variety, are generally available from only one or two supplier and one or two qualities of each item offered for sale. Canned and frozen foods may offer several labels, but with very similar products inside the can or box. The variety of different items offered by supermarkets is enormous, but efficiency dictates that substantive choices within generic food groups remain very limited. The emergence of fast food restaurants, likewise, has transformed table-ready beef, chicken and potatoes into a dozen or so manufactured, industrial commodities that can be mass produced for mass consumption. People who are satisfied with products that fit industrial systems of mass production realize a bargain. They get what they want at a lower price. Those who are not satisfied, but just go along, represent potentially profitable niche markets.

Markets for value-added agricultural products are often confused with niche markets. However, most value-added items in supermarkets are more characteristic of mass production than are the basic commodities from which they are made. The markets for services that result in convenience and easy preparation are far larger and more generic that are the markets for beef or broccoli. Those services are mass produced in large, specialized processing firms and are promoted through multimillion dollar advertising budgets. It might be more logical for an individual farmer to compete with IBP in processing beef than to compete with Banquet Foods in producing TV dinners. Many other value-added activities are equally price competitive. Farmers who attempt to add value through processing, transportation, or storage must be willing and able to perform those functions better or at a lower costs than can existing marketing firms, if they expect to make a profit. Even if such farmers are successful, they are not engaging in niche marketing. Instead, they have become successful players in the mass marketing game.

Markets of the Future

The conventional wisdom among those in the agricultural establishment is that trends toward industrial production for mass markets are trends of future for American agriculture. Agriculture is becoming just another industrial sector of our industrial economy. But the world is continually changing. A growing number of people who make their living forecasting the future, the futurist, see a new, post-industrial era where there is room for more than one model or paradigm for economic and human progress (Toffler, 1990; Drucker, 1989; Naisbitt and Aburdene, 1990; Reich, 1992, to name a few).

Noted futurist Alvin Toffler, in his book Powershift, points out that many forecasters simply present unrelated trends, as if they would continue indefinitely, without providing any insight regarding how the trends are interconnected or the forces likely to reverse them. He contends the forces of industrialization have pretty well run their course and already show signs of reversing. He labels the industrial models of economic progress as becoming increasingly obsolete. He claim that old notions of efficiency and productivity are no longer valid. Mass production is no longer a symbol of "modern" business operation. The new "modern" model is to produce customized goods and services aimed at niche markets, to constantly innovate, to focus on value-added products and tailored production.

He goes on to state "the most important economic development of our lifetime has been the rise of a new system of creating wealth, based no longer on muscle but on the mind" (Toffler, p. 9). He contends that "the conventional factors of production -- land, labor, raw materials, and capital -- become less important as knowledge is substituted for them" (Toffler, p. 238). "Because it reduces the need for raw material, labor, time, space, and capital, knowledge becomes the central resource of the advanced economy (Toffler, p. 91).

Toffler also provides some insights into the nature of knowledge-based production. He states that separate and sequential systems of production are being replaced with synthesis and simultaneous systems of production. Synergism is replacing specialization as a source of production efficiency. Tailoring products to desires of specific customers is replacing low price as the source of value. Simultaneity, synthesis, synergism, tailored production -- this is the "mind work" of the future.

Peter Drucker, a noted business consultant and author, talks of the "Post Business Society" in his book, The New Realities. He states "the biggest shift -- bigger by far than the changes in politics, government or economics -- is the shift to the knowledge society. The social center of gravity has shifted to the knowledge worker. All developed countries are becoming post-business, knowledge societies. Looked at one way, this is the logical result of a long evolution in which we moved from working by the sweat of our brow and by muscle to industrial work and finally to knowledge work" (Drucker, 1989, p. 173).

Drucker contends that there is an important, fundamental difference between knowledge work and industrial work. Industrial work is fundamentally a mechanical process, whereas the basic principle of knowledge work is biological. He relates this difference to determining the "right size" of organization required to perform a given task. "Greater performance in a mechanical system is obtained by scaling up. Greater power means greater output: bigger is better. But this does not hold for biological systems.

There, size follows function. It would surely be counterproductive for a cockroach to be big, and equally counterproductive for the elephant to be small. As biologists are fond of saying, "The rat knows everything it needs to know to be a successful rat." Whether the rat is more intelligent than the human being is a stupid question; in what it takes to be a successful rat, the rat is way ahead of any other animal, including human beings" (Drucker, 1989, p. 259).

Differences in organizing principles may be critically important in determining the future size and ownership structure of economic enterprises, including farms. Other things equal, the smallest effective size is best for enterprises based on information and knowledge work. "'Bigger' will be 'better' only if the task cannot be done otherwise" (Drucker, 1989 p. 260).

Some see the future knowledge society as one where large industrial firms will continue to dominate. However, Drucker points out that "in the knowledge society into which we are moving, individuals are central. Knowledge is not impersonal , like money. Knowledge does not reside in a book, a data bank, a software program; they contain only information. Knowledge is always embodied in a person; carried by a person; created, augmented, or improved by a person; used or misused by a person. n The shift to the knowledge society therefore puts the person in the center" (Drucker, 1993, p. 210). Henry Ford is quoted as saying that his biggest problem was that he had to hire "whole people" when all he needed was their "two hands." Knowledge work and industrial work are fundamentally incompatible.

The significance of the above quotes is not that they reflect some inherent truth, but rather their consistency with the paradigm of sustainable agriculture. A sustainable agriculture will require site-specific, individualistic, dynamic, management-intensive, knowledge-base systems of production and distribution. Sustainability will require a matching of size and function reflected in a variety of sizes and types of agricultural enterprises to sustain a productive natural resource base, a healthy economy, and a progressive human society. The emerging paradigm for a sustainable agriculture is fundamentally consistent with the emerging post-industrial paradigm for human progress.

Niche Marketing for Sustainability

The key to successful niche marketing is to find a market that is: (a) sufficiently different from its mass market context to allow a significant premium in price (or a significantly lower cost), (b) large enough to be served profitably, but (c) too small to accommodate methods of mass production and distribution. Industrial suppliers attempt to fill market niches by providing as wide a variety of products and services as they can economically provide. However, much of their differentiation is in packaging, convenience, and other largely superficial differences in the same basic products. The primary advantage of being big arises from the ability to specialize, mechanize, routinize, and realize the economies of scale of "mass" production methods. Whenever industrial suppliers attempt to target niche markets, their production methods must become more like those of small suppliers, and consequently they become more vulnerable to small-firm competition. The smaller the market niche, the greater the competitive advantage for a similarly small supplier.

The basic market assumption of the industrial paradigm is "the consumer is always right." The industrial economy is "consumer driven" in that successful producers must fulfill the needs and wants of consumers. In reality, consumers are considered to be right only if their choices accommodate the needs of the industrial economy. If consumers cannot be bribed, persuaded, or coerced to conform to the requirements of mass production, they are ignored by industrial suppliers. These ignored consumers represent potentially profitable niche markets.

The basic market assumption of the sustainable paradigm is "the needs and wants of both consumers and producers must accommodate long run ecologic, economic, and social sustainability." Niche marketing motivated solely, or mostly, by the desire to increase profits may contribute no more to long run sustainability that does any other form of market segmentation or product differentiation (DeLind, 1994). The role of niche marketing in sustainability is to enhance the economic viability of systems that are ecologically sound and socially responsible and otherwise likely to be sustainable.

Sustainable niche marketing requires a matching of resource management, production, marketing, and consumption in ways that sustain the whole vertical system. No one element of that vertical system is considered supreme, neither consumer nor producer. People must protect and sustain the resource base, but the resource base must support and sustain human society. The economic system must provide the incentives and means for sustaining both resources and people. The diversity in size and type of economic enterprises must match the diversity within the natural resource base and diversity within human society.

If the road to agricultural sustainability leads toward larger numbers of smaller, more diversified farms, then niche marketing represents an opportunity for smaller, diversified farms to expand vertically and to compete commercially with larger, specialized agricultural enterprises. Sustainable niche markets that grow into mass markets will remain sustainable only if the production systems that support them remain sustainable. However in the post industrial era of human progress, niche markets may well become the norm rather than exception. The natural resource base is inherently diverse as are the abilities and aspirations of producers and the tastes and preferences of consumers. A similarly diverse collection of diverse small farms and matching niche markets might logically characterize an agriculture capable of meeting the needs and wants of people by means that are sustainable. Niche marketing may well be a critical key to long run agricultural sustainability.

4

GOVERNMENT PROGRAMMES AND REGULATIONS FOR AGRIBUSINESS

The welfare of farmers has always been the top priority of the Government of India. For this, it has implemented different schemes or yojana's to revive the agriculture sector and to improve the economic conditions of farmers. These agricultural schemes are very beneficial for the farmers and they must know about it so as to take its benefit. So through this article, we will tell you about some of the most useful and popular government schemes for farmers in India.

GOVERNMENT PROGRAMMES FOR AGRIBUSINESS

Both the central government and state governments have devised programmes to woo the entrepreneurs for setting up agribusiness in India. They may be classified as programmes falling under small-scale industries, Khaki village industries, small and medium industries, large industries based on the finance required for investment. Apart from these to harness the export market after signing the WTO agreement. They include conversion of Export Processing Zones (EPZ) into Special Economic Zones (SEZ), establishment of Agra-export zones. For which government is attracting private investors to make investment in infrastructure development like cold storage chains, improvement in road, rail, sea and air transport systems.

The Ministry of Small Scale Industries & Agro and Rural Industries designs and implements the policies through its field organizations for promotion and growth of small and tiny enterprises, including the coir industries. The Ministry also coordinates with other Ministries / Departments on behalf the Small Scale Industries (SSI) sector.

The implementation of policies and various programmes/schemes for providing infrastructure and support services to small enterprises is undertaken through its attached office, namely the Small Industry Development Organization (SIDO), statutory bodies/other organizations likely Khaki and Village Industries Commission (KVIC) & Coir Board a Public Sector Undertaking - National Small Industries Corporation (NSIC) and three training institutes - National Institute of Small Industry Extension Training (NISIET), Hyderabad, National Institute for Entrepreneurship and Small Business Development (NIESBUD), New Delhi and Indian Institute of Entrepreneurship (IIE), Guwahati.

Entrepreneurial Training institutes

There are three institutes engaged in training of small-scale entrepreneurs. These are Indian Institute of Entrepreneurship (IIE), Guwahati, National Institute of Small Industry Extension Training (NISIET), Hyderabad, National Institute for Entrepreneurship and Small Business Development (NIESBUD), New Delhi.

Export Processing Zones

The export zones (EPZ) set up as enclaves, separated from the Domestic Tariffs Areas by fiscal barriers, are intended to provide a competitive duty free environment for export production. There are four EPZs set up by the Government at Naiad (Uttar Pradesh), Chennai (Tamil Nadu), Plata (West Bengal) and Vishakhapatnam (Andhra Pradesh).

Special Economic Zones

A new scheme for setting up of Special Economic Zones (SEZs) in the country to promote exports was announced by the Government in the Export and Import Policy on 31st March, 2000. The policy provided for setting up of SEZs in the public, private, joint sectors or by State Governments. It was also announced that some of the existing Export Processing Zones would be converted into SEZs. Accordingly, the Government has issued notification on 1-11-2000 for conversion of the existing Export Processing Zones at Kendal (Gujarat), Santa Cruz (Maharashtra) and Cochin (Kerala) into SEZs. Notification has also been issued for conversion of the private sector EPZ at Seurat (Gujarat) into the Special Economic Zone at the request of the promoters.

Agri-Business Centres

Agri-Business Centres are commercial units of agri-ventures established by trained agriculture professionals. These ventures may include maintenance and custom hiring of farm equipment, sale of inputs and other services in agriculture and allied areas, including post-harvest management and market linkages for income generation and entrepreneurship development.

The scheme covers full financial support for training and handholding, provision of loan and credit-linked back-end composite subsidy.

Export Oriented Units (EOU)

The export Oriented units (EOU) scheme introduced in the early 1981, is complementary to the EPZ scheme. It adopts the same production regime but offers a wide option in locations with reference to factors like source of materials, ports of export, hunter land facilities, and availability of technological skills, existence of industrial base and the need for a large area of land for the project. And 1,536 units are in operation under the EPU scheme as on March, 2001.

Product range: EOUs are mainly concentrated in textiles and yarn, food processing, electronics, chemicals, plastics, granites and minerals/ores. Majority of units are located in Tamil Nadu, Andhra Pradesh, Karnataka, Maharashtra and Gujarat.

Export Promotion Industrial Park (EPIP) Scheme: A centrally sponsored "Export Promotion Industrial Park (EPIP) Scheme has been introduced in 1993-94 with a view to involving the State Governments in the creation of infrastructure facilities for export oriented production. The scheme provides that 75 % of the capital expenditure incurred towards creation of such facilities, ordinarily limited to Rs. 10 cores in each case, will be met from a central grant to the State Governments.

New Anna Marumalarchi Theta: The State Government of Tamil Nadu has introduced the New Anna Marumalarchi Theta in April 2002 to set up agribusiness units with a minimum investment of Rs 35 laths at the rate of one unit in each block. This scheme provides scope for setting 385 agribusiness units in Tamil Nadu.

Agra-Clinic and Agribusiness Centres: Small Farmers Agribusiness Consortium in co-operation with MANAGE has drawn plans to provide training on management capacity building for those willing to set up Agriclinics and Agribusiness Centres either as individual or a group five (four agricultural and allied graduates and one management graduate) with a maximum loan assistance of Rs. 10 laths for individuals and Rs 50 laths for a group of five entrepreneurs.

REGULATIONS

Legal regulations

Decisions are strongly affected by laws pertaining to competition, price setting, distribution arrangements, advertising, etc. It is necessary for a manager to understand the legal environment of the country and the jurisdiction of its courts. The following laws affecting business in India are important.

1. Indian Contract Act, 1872
2. Factories Act, 1948
3. Minimum Wages Act, 1948
4. Securities contracts Regulation Act, 1956 (Now replaced by SEBI Act)
5. The Companies Act, 1956
6. Trade and merchandise Marks Act, 1958
7. Monopolies and Restrictive Trade Practices Act, 1969
8. The water (Prevention and Control of Pollution) Act, 1974
9. The Air (Prevention and Control of pollution) Act, 1981
10. Sick Industrial Companies (Special provision) act, 1985
11. Environment protection Act, 1986
12. Consumer protection Act, 1986
13. Securities and Exchange Board of India Act, 1992
14. Taxation laws covering Corporate tax, indirect taxes like Excise, Customs, Sales tax and Wealth tax)

Schemes introduced by the government to make agriculture a profitable sector:

The agriculture sector contributed 51.9 percent to India's GDP in 1950. Since then it has been on a downside and it currently stands at 13.9 percent. However, a change from an agrarian-centric economy to an industry-centric economy is inevitable with the advent of industries. With industries growing at a faster pace than the rate at which trees are being planted, will there be a time when agriculture's productivity dwindles to a null? If yes, is it already here?

Living in a country where the cattle are worshipped as a goddess, about 60 percent of the population was banking on agriculture for their main source of income during the 1950s. Despite half of the population still continuing with the profession, the returns are low. While urbanisation might be cited as a reason, it is hard not to neglect the fact that agriculture is no more a profitable sector. Infrastructure costs have started running high, with its maintenance cost and capital investment only adding on to the farmers' misery. According to an article by The Hindu, the average recovery rate of the investments made by Indian farmer is only 30 percent.

Another cause for low productivity is small holdings of land with farmers. By owning a fragmented land, effective irrigation and optimum usage of fertilisers for crops becomes difficult, thus resulting in lower yields. In India, more than two-thirds of the crops lack proper irrigational facilities, albeit India being the second largest irrigated country after China. But improper irrigation can also lead to other problems affecting yield like soil erosion, salinity, etc.

In the wake of agriculture losing its lucrative appeal, budget 2016-17 has proposed to bring 2.85 million hectares under irrigation, Rs 2,87,000 to be donated and 100 percent electrification to all villages by May 2018. The government has also announced a couple of initiatives to resolve the farmers' plight.

Some of the recent developments in the agriculture and allied sector are enumerated below:

Launch of Pradhan Mantri Fasal Bima Yojana

Farming has become an unreliable sector. Farmers are always unsure of the yield they'll reap, but strive to draw the maximum benefits out of their investments and effort. Often farmers might be at the receiving end, with natural calamities like droughts and floods affecting their yield adversely. To resolve the problem of unpredictable nature of farming and prevent farmer suicides in the country, the Government launched Pradhan Mantri Fasal Bima Yojana in early 2016. It's a crop insurance policy with relaxed premium rates on the principal sum insured for farmers. Implemented with a budget of Rs 17,600 core, this scheme will provide financial support to farmers and cover for their losses.

After green, white, and golden, it's time for blue

The Cabinet Committee on Economic Affairs (CCEA) has approved Blue Revolution in India. It's an integrated scheme designed to increase the productivity and profitability from aquaculture and fisheries resources, inclusive of both inland and marine. With a budget of Rs 3,000 core offered by the government for the next five years, this scheme aims to maintain an annual growth rate of six to eight percent of the agriculture and allied sector.

Government to invest Rs 221 core to improve milk productivity

India boasts of being the largest producer of milk in the world with an annual output of 130 million tonnes. However, with a milk-producing animal population of more than 118 million, the milk yields per animal is very low. To meet the steadily growing demand for milk, the National Dairy Development Board (NDDB) has announced 42 dairy projects, under a budget of 221 cores. These projects shall focus on improving the milk productivity of major milk-producing states like Uttar Pradesh, Maharashtra, Karnataka, Tamil Nadu and the likes.

Energy-efficient irrigation to be implemented

A report says that in India more than two-thirds of the arable area lacks proper irrigational facilities. Taking note of this, Power Minister Pixyish Goal said that the government is planning on investing Rs 75,000 core to provide energy-efficient irrigational facilities to farmers, over the next three to four years. Under this scheme, close to 30 million energy-saving pump sets would be given to farmers and this cost would be recovered via savings in the electricity consumed. This would result in about 46 billion kWh of power being saved and creation of 20 laky jobs.

Launch of Paramparagat Krishi Vikas Yojana

The government has launched Paramparagat Krishi Vikas Yojana in order to address the critical importance of soil and water for improving agricultural production. The government would support and improve the organic farming practices prevalent in India. Following cluster approach mode of farming, at least 50 farmers would form a group having 50 acres of land to implement organic farming. The government aims to cover 10,000 clusters and five laky hectares of arable land under organic farming within three years.

Recently, the government has been active in investing in agricultural infrastructure such as irrigational facilities, mechanised farming, and warehousing. The growing use of genetically modified crops will also improve the sector's contribution to GDP. While all of these initiatives look promising, in what way are they going to affect the current scenario are something interesting to watch out for.

GOVERNMENT SUBSIDY SCHEMES FOR AGRICULTURE IN INDIA:

Basically, agriculture or farming is an evergreen and sunshine sector. Not only the farmers, the subsidy schemes are beneficial for Argil entrepreneurs, corporations, SHG Groups and farmers' cooperatives also. Additionally, in some cases, the sharecroppers and tenant farmers can also avail the agriculture subsidy through some particular schemes.

We have compiled a list of 10 best Government subsidy schemes for agriculture. In addition to that, we have described the eligibility criteria and who can apply for the scheme.

List of 10 Best Government Subsidy Schemes for Agriculture

1. Rural Go down Scheme

The rural go down scheme comes under the schemes related to agriculture marketing. Basically, the scheme promotes the grading, standardization and quality control of agricultural products to improve their marketability. In addition, the scheme helps in prevention of distress sale by providing the facility of pledge financing and marketing credit.

Who Can Apply

- Farmers group (Commodity Interest Group)
- Farmers Federation
- Argil. Entrepreneur /Individual farmer
- Partnership/ proprietary firms
- Farmer's cooperatives

2. Development/Strengthening of Agricultural Marketing Infrastructure, Grading and Standardization

The scheme also comes under the Agriculture Marketing scheme. Basically, the scheme promotes direct marketing so as to increase market efficiency through reduction in intermediaries and handling channels. Therefore, it helps in enhancing farmers' income.

Who Can Apply

- Individual farmer
- Group of farmers/ growers/
- Partnership/ Proprietary firms
- Non-Government Organizations (NGOs)
- Self Help Groups (SHGs)
- Companies, Corporations

3. Modified National Agricultural Insurance Scheme (MNAIS)

Actually, MNAIS comes under the credit related subsidy schemes. Generally, the scheme provides insurance coverage and financial support to the farmers. It provides financial assistance in the event of failure of any of the notified crop as a result of natural calamities, pests & diseases.

Who Can Apply

- All farmers including sharecroppers and tenant farmers
- Individual owner-cultivator
- Farmers enrolled under contract farming, directly or through promoters/organizers
- Groups of farmers/societies serviced by Fertilizer Companies
- Pesticide firms, Crop Growers associations, Self Help Groups (SHGs), Nongovernmental

Organizations (NGOs)

4. Coconut Palm Insurance Scheme (CPIS)

Basically, CPIS assists the coconut growers in insuring coconut palms, against natural and other perils. In addition, it provides timely relief to farmers, who suffer income loss due to a sudden death of palms.

Who Can Apply

Individual farmer/grower offering at least 10 healthy nut bearing palms for insurance in specified age group, (4-60 years for dwarf, hybrid and 7-60 years for tall) in contiguous area/plot are eligible for insurance

5. Central Sector Scheme Establishment of Agri-Clinics & Agri-Business Centres (ACABC)

The Central Sector Scheme "Establishment of the Agri-clinics and Agri-business Centres (ACABC)" is under implementation since 2002. This program aims to tap the expertise available in the large pool of Agriculture Graduates. Irrespective of whether you are a fresh graduate or not or whether you are currently employed or not, you can set up your own AgriClinic or Agribusiness Centre and offer professional extension services to innumerable farmers.

Who Can Apply

- Argil Entrepreneur
- Individual farmer

6. National Horticultural Mission

National Horticulture Mission (NHM) is one of the sub schemes of Mission for Integrated Development of Horticulture (MIDH). Additionally, the schemes provide support to enhance horticulture production, improve nutritional security and income support to farm households.

Who Can Apply

- Farmer's Association
- Farmers' Companies
- Registered Farmers' Societies
- Farmers' Cooperatives
- Extension workers

7. *Aspire Scheme*

The Government of India has launched the ASPIRE scheme to promote the rural agro-industry in India. The scheme is for the promotion of innovation, entrepreneurship, and agro-industry. The proposed scheme is designed to provide the necessary skill set for setting up business enterprises and also to facilitate the market linkages available to the entrepreneurs.

Who Can Apply

- Individual farmers
- Argil Entrepreneurs

8. *National Bamboo Mission*

Basically, the National Bamboo Mission provides the growth of the bamboo sector through as an area based regionally differentiated strategy. Additionally, the scheme promotes marketing of bamboo and bamboo based handicrafts.

Who Can Apply

- Farmer's Association
- Farmers' Companies
- Bamboo growers
- Extension workers

9. *National Project on organic farming (NPOF)*

Basically, the NPOF promotes the organic farming practice. Additionally, it helps to popularize organic farming for enhancing farm income. Also, it promotes Organic Farming through low-cost certification system known as "Participatory Guarantee System".

Who Can Apply

- Individual Farmer
- Farmers' group

- Agra- entrepreneur
- Service provider

10. Sub-Mission on Seed and Planting Material (SMSP)

Actually, the SMSP aims to development and strengthening of infrastructure facilities for production and distribution of quality seeds (DSIS). Additionally, the scheme encourages the seed treatment, particularly for farm saved seed. Furthermore, it promotes new technologies and methodologies in seed production, processing, testing etc.

Who Can Apply

Individual farmers

Agriculture and allied sector account for 14% of the country's GDP and 13% of overall exports. In addition, the sector provides employment to over 50% of the work force in the country. However, the agriculture and farming activity involves risk of damage and loss. Additionally, it demands both short term and long term capital investment. We hope, this list of best subsidy schemes for agriculture will help you in getting the financial assistance in your particular project.

Schemes

This section provides you useful links to avail various Citizen Services being provided by the Central and State/UT Governments in India. The list, however, is not exhaustive, as we are committed to adding more and more information about other services for which citizens and other stakeholders need to interact with the Government.

Schemes related to central ministries

Central

- National Bio-Science awards for career development
- Agricultural Census
- Jute Technology Mission
- Technology Mission on Cotton
- Rashtriya Krishi Vikas Yojana (RKVY)
- Livestock Insurance Scheme
- National Scheme on Welfare of Fishermen
- Scheme on Fisheries Training and Extension

- Gramin Bhandaran Yojana
- Central Assistance to State Plan Scheme on Watershed Development
- Capacity Building to enhance Competitiveness of Indian Agriculture and Registration of Organic Proud

Schemes related to state departments

State	Schemes
Andaman and Nicobar Island	Supply of inboard/outboard engines at 50 per cent subsidy Fodder Development Programme
Andhra Pradesh	Polambadi Seed Village Scheme Rashtriya Krishi Vikas Yojana Economic Assistance (Bankable Scheme) Scheme of Development of Primitive Tribal Groups National Horticulture Mission
Chhattisgarh	Chatragrih Yojana Book Bank Yojana
Goa	Development of Agricultural Extension Development of Manure Scheme Soil and Water Conservations Scheme National Horticulture Mission Scheme Cultivation of Red Oil Palm (ISOPOM) Scheme Agricultural Research & Education
Gujarat	Scheme for Integrated Horticulture Development Programmes
Himachal Pradesh	Flood Protection Works Command Area Development Programme Quality Seed Multiplication and Distribution Crop Improvement Programme Rashtriya Krishi Bima Yojana

	Rural Infrastructure Development Fund (RIDF) Prime Minister Rojgar Yojana(PMRY) Entrepreneurial Development of Women Farmers Development of Beekeeping Establishment of Tissue/Leaf Analysis Laboratory Establishment of Plant Health Clinic Incentive to Farmers MT Non-Agricultural Loan Scheme Society Agriculture Loan Scheme MT Agriculture Loan Scheme Rural Development through Diversification in Agriculture Self Reliance through Sericulture and Dairy Development Kisan Credit Card Scheme Subsidy on Horticulture Production Inputs to Small Farmers General Loan Scheme Him Swalamban Yojana-Loan Scheme for SC/ST Persons for Starting Transport, Service, Agriculture Sector
Haryana	Scheme for Agriculture Engineering and Services Soil Conservation Sustainable Development of Sugarcane Based Cropping System (SUBACS) Macro Management of Agriculture Financial Assistance to Panchayats and Panchayat Semites
Jammu and Kashmir	Agriculture Export Zones Technology Mission for Post Harvest Management, Technology Transfer Marketing and Exports
Kerala	Scheme for Strengthening of Agricultural Extension Rashtriya Krishi Vikas Yojana (RKVY) Commodity Safety Net Scheme Development of Vegetable Cultivation in Villages

	Crop Insurance Scheme Scheme for Integrated Development of Coconut Holdings in the State Sustainable Development of Rice Based Farming System Schemes for Integrated Pest Management System Modernisation of Agricultural Farms Scheme for Soil and Water Conservation on Watershed Basis National Agricultural Insurance Scheme (NAIS) Scheme to Expand Information Services for Farm Information and Communication in the State Integrated Development of Scampi/Fish Farming Small Farm Mechanisation
Maharashtra	Tree Planting on Community Lands in Identified Watersheds Western Ghats Development Programme NABARD-WDF Assisted Watershed Development Programme River Valley Project Gram Sachivalaya Yojana (GSY)
Meghalaya	Sampoorna Gramin Rojgar Yojana (SGRY)
Manipur	Pradhan Mantri Gram Sadak Yojana (PMGSY) Macro Management Mode of Agriculture Oilseed Production Programme National Pulse Development Programme Accelerated Maize Development Programme(AMDP) Integrated Cereal Development Programme (ICDP)-Rice Sustainable development on sugarcane base cropping system(SUBACS) Strengthening of Existing Laboratory

	Farm Mechanization Rodent Control Integrated Pest Management Reclamation of water logged lands Promotion of Green Manure Crops Agriculture Technology Management Agency (ATMA) Watershed Development Project in Shifting Cultivation Areas (WDPSCA) Rashtriya Krishi Vikas Yojana (RKVY)
Madhya Pradesh	Krishi Mitra Yojana Sooraj-Dhara Yojana Bhoo-Jal Samwardhan Yojana Rashtriya Jalgrahan Koshal Vikas Karyakaram Rashtriya Krishi Bima Yojana Ponds Construction Scheme for Irrigation Rashtriya Biogas Vikas Yojana Balsam Tail Construction Scheme Annapurna Yojana Sagan Kapuas Vikas Isopod Yojana Sugarcane Development Scheme
	Unified Grains Development Scheme Ashtray Kristi Bema Yana Minor Irrigation Scheme Amalgamated Special Fund (ASF) Schemes River Bank/ Flood Water conservation Scheme Micro Management of Agriculture Laugh Sinclair Yojnantargat Nalco Khan an Yana
Mizoram	National Agricultural Insurance Scheme Farm Mechanization

Odessa	Scheme On Consumer Clubs Prada Mantra Gram Sadat Yana, PMGSY Watershed Development Programme Rashtriya Krishi Vikas Yojana National Project on Management of Soil Health and Fertility National Food Security Mission Command Area Development & Water Management (CADWM) Programme Scheme for Pain Panchayat Scheme on Organic Farming Integrated Rural Energy Planning (IREP) Orissa Tribal Empowerment and Livelihood Programme(OTELP) Component Wise Operational Guideline for ISOPOM 2007-08 Integrated Scheme of Oilseeds, Pulses & Maize Farm Mechanization Scheme Balance and Integrated use of Fertiliser Scheme Sugarcane Development Scheme
Punjab	Subsidy on Recoil/Hybrid seeds/ Demonstration plots Prime Minister's Rojgar Yojana (PMRY) Anaya Magritte Jota Scheme Training Infrastructure in Dairying for Rural Dairy Farmers Farmer's Welfare Scheme Rashtriya Krishi Vikas Yojana (RKVY) Mai Bag Istria Shakhty Scheme
Pondicherry	Schemes for Agricultural Workers Welfare Society Key Village Scheme

Rajasthan	Doctor Ka Ana Clinic Scheme, 2008 Women Development Project Land Purchase Scheme for Scheduled Caste/Tribe Women Integrated Horticulture Development Scheme Schemes under Tamil Nadu Watershed Development Agency (TAWDEVA) Tamil Nadu Agricultural Labourers - Farmers (Social Security and Welfare) Scheme, 2006
Tripura	Sampuran Gramin Rojgar Yojana (SGRY)

AGRICULTURE SCHEMES:

A. Krishonnati Yojana

It is an umbrella scheme which includes the following:

- National Food Security Mission (NFSM): to increase the production of rice, wheat and pulses. The mission is being continued during 12th plan with new target of additional production of 25 million tonnes of food grains comprising 10 million tonnes of rice, 8 million tonnes of wheat and 4 million tonnes of pulses and 3 million tonnes of coarse cereals.
- National Food Security Mission-Commercial Crops: for crop development programme on cotton, jute and sugarcane for enhancing productivity
- Mission for Integrated Development of Horticulture (MIDH): It covers wide horticulture base, which includes fruits, vegetables, tuber crops, mushrooms, spices and aromatic plants flowers and foliage and plantation crops like coconut, areca nut, cashew nut, cocoa and bamboo.
- National Mission on Oilseeds and Oil Palm: envisages increase in production of vegetable oils sourced from oilseeds, oil palm and tree borne oilseeds.
- National Mission for Sustainable Agriculture: aims at making agriculture more productive, sustainable and remunerative and climate resilient by promoting location specific integrated/composite farming systems; soil and moisture conservation measures; comprehensive soil health management; efficient water management practices and mainstreaming rained technologies.
- National Mission on Agricultural Extension and Technology: Its aim is to restructure and strengthen agricultural extension to enable delivery of

appropriate technology and improved agronomic practices to the farmers consist.

B. Initiative for increasing flow of credit

- In order to ensure that all eligible farmers are provided with hassle free and timely credit for their agricultural operation, Kinas Credit Card (KCC) Scheme was introduced in 1998-99. The main objectives of the scheme are to meet the short term credit requirements for cultivation of crops, post harvest expenses, produce marketing loan, consumption requirements of farmer household, working capital for maintenance of farm assets and activities allied to agriculture like dairy animals, inland fishery, etc., investment credit requirement for agriculture and allied activities like pump sets, sprayers, dairy animals, etc.

C. Pradhan Mantri Fasal Bima Yojana

- Under the new scheme, the farmers' premium has been kept at a maximum of 2 per cent for food grains and oilseeds, and up to 5 per cent for horticulture and cotton crops.
- There is no upper limit on Government subsidy. Even if balance premium is 90%, it will be borne by the Government. Earlier, there was a provision of capping the premium rate which resulted in low claims being paid to farmers. This capping was done to limit Government outgo on the premium subsidy. This capping has now been removed and farmers will get claim against full sum insured without any reduction.
- Importantly for the beneficiaries, crop losses which are covered under the scheme include Yield Losses as well as post-harvest losses, where coverage will be available up to a maximum period of 14 days from harvesting for those crops.
- The use of technology will be encouraged to a great extent resulting in operational efficiency. Smart phones will be used to capture and upload data of crop cutting to reduce the delays in claim payment to farmers. Remote sensing will be used to reduce the number of crop cutting experiments.

D. Mere Gain, Mere Aura

- this scheme is being launched involving agricultural experts of agricultural universities and ICAR institutes for effective and deeper reach of scientific farming to the villages.
- A group of experts will be associated with one particular village to create awareness and adoption of new technologies including farm investment, loans, availability of inputs and marketing.

- All the scientists from ICAR and agricultural universities will participate in this initiative.

E. Kristi Dark

- IARI initiated this novel scheme in 20 districts in which postmen supplied seeds of improved varieties of crops to the farmers in far-flung areas.
- Owing to its success and popularity, this scheme is being extended in 100 districts of 14 states with the association of Krishi Vigyan Kendra's.
- This will provide improved seed to farmers at their doorstep.

F. Soil Health Card

- Soil Health cards are necessary to ensure that only requisite nutrients are applied in the soil in a balanced manner to enhance productivity of specific crops in a sustainable manner.
- Values on soil parameters such as pH, EC, N, P, K, S, Zn, Fe, Man, Cu & B.
- Recommendation on appropriate dosage of fertilizer application based on test values and requirement of crop, use of organic manures and soil amendments to acidic/alkaline/sonic soils.

G. Paramparagat Krishi Vikas Yojana(PKVY)

- Aim of the project is to maximize the utilization of natural resources through eco-friendly cultivation.
- Organic farming is a method of farming system which primarily aimed at cultivating the land and raising crops in such a way, as to keep the soil alive and in good health by use of organic wastes (crop, animal and farm wastes, aquatic wastes) and other biological materials along with beneficial microbes (bio-fertilizers) to release nutrients to crops for increased sustainable production in an eco friendly pollution free environment.

H. Promotion of National Market through Agra Tech Infrastructure Fund (ATIF)

- Central Sector Scheme for Promotion of National Agricultural Market through Agra-Tech Infrastructure Fund (ATIF) for Rs.200 cores to be implemented during 2014-15 to 2016-17.
- The Scheme envisages initiation of e-marketing platform at the national level and will support creation of infrastructure to enable e-marketing in 642 regulated markets across the country.

- For creation of a National Market, a common platform across all States is necessary. It is, therefore, proposed that a Service Provider be engaged centrally who would build, operate and maintain the e-platform on PPP (Build, Own, Operate, Transfer - BOOT) model. This platform would be customized/ configured to address the variations in different states.
- As an initiative of deregulation, States have been advised by the Government of India to bring fruits and vegetables out of the ambit of APMC Act. In pursuance of this advisory, 12 States have, so far, either de-regulated the marketing of fruits and vegetables or have exempted from levying of market fee.

I. MUDRA Bank

- The Finance Minister has proposed to create a Micro Units Development Refinance Agency (MUDRA) Bank, with a corpus of Rs. 20,000 core, and credit guarantee corpus of 3,000 core, which will refinance Micro-Finance Institutions through a Prada Mantra Madras Yolanda.
- will be given to SC/ST enterprises in lending. MUDRA Bank will operate through regional level financing institutions who in turn will connect with last mile lenders such as MFIs, Small Banks, Primary Credit Cooperative Societies, Self Help Groups (SHGs), NBFC (other than MFI) and other lending institutions.
- MUDRA Bank will refinance Micro-Finance Institutions through a Prada Mantra Madras Yolanda (PMMY). In lending, priority will be given to SC/ST enterprises. These measures will greatly increase the confidence of young, educated or skilled workers who would not be able to aspire to become first generation entrepreneurs; existing small businesses, too will be able to expand their activities. Since the MUDRA Bank will be set up through an enactment of law and it will take some time.

J. Kristi Kalian Cess:

- Union Budget for 2016-17 (April-March) introduced a new cess on services named Kristi Kalian Cess at the rate of 0.5%. The effective rate of the Kristi Kalian Cess, however, will be lower than 0.5% as the government will provide input tax credit for the cess, as against no input tax credit for Swatch Bharat Cess.
- Proceeds of cess would be exclusively used for financing initiatives relating to improvement of agriculture and welfare of farmers.

K. Direct Benefit Transfer (DBT) For Fertilizer Sector:

- Government also announced to introduce Direct Benefit Transfer of fertilizer subsidy to farmers on pilot basis in few districts of the country.
- The government has fixed the Maximum Retail Price (MRP) of Urea at Rs. 5,360 per tones.
- The government had allocated Rs.72, 968.56 cores for fertiliser subsidy, out of which Rs. 38,200 core was earmarked for domestic Urea.

PROGRAMMES, SCHEMES & NEW INITIATIVES:

National Mission for Sustainable Agriculture (NMSA) The National Mission for Sustainable Agriculture (NMSA), which is one of the eight missions under the National Action Plan on Climate Change (NAPCC), seeks to address issues associated with climate change. Adaptation and mitigation strategies need to address food security, equitable access of food resources, enhancing livelihood opportunities and contributing to economic stability of the people at the end. Mission, therefore, focuses to transform Indian agriculture into a climate resilient production system through suitable adaptation and mitigation measures mainly in the domain of crops and animal husbandry. Mission interventions are judiciously embedded in research and development activities, absorption of improved technology and best practices, creation of physical and financial infrastructure and institutional framework, facilitating access to information and capacity building. While dry land agriculture will receive focused importance by way of developing suitable drought and pest resistant crop varieties with necessary institutional support, the mission would also expand its coverage to rained areas for integrating farming systems in farms with local agro ecology and also in management of livestock and fisheries, so that the agricultural production system continues to grow in a sustainable manner on one hand and natural resources are also conserved and utilized.

NMSA as a programmatic intervention made operational from the year 2014-15 aims at making agriculture more productive, sustainable, and remunerative and climate resilient by promoting location specific integrated/composite farming systems; soil and moisture conservation measures; comprehensive soil health management; efficient water management practices and mainstreaming rained technologies.

NMSA has the following major components

Rained Area Development

- Integrated Farming System (IFS) is being promoted under RAD in which activities like horticulture, livestock, fishery, agro forestry, value addition are to be taken up along with crops/cropping system.

- SHM is aimed at promoting location as well as crop specific sustainable soil health management including residue management, organic farming practices by way of creating and linking soil fertility maps with macro-micro nutrient management, appropriate land use based on land type.

Implementing Divisions

- SHM component is being implemented by INM Division...
- RAD component is being implemented by RFS Division.

Rained Area Development (RAD) as a component of NMSA is being implemented in the country from 2014-15. The RAD focuses on Integrated Farming System (IFS) for enhancing productivity and minimizing risks associated with climatic variability's. Under this system, crops/cropping system is integrated with activities like horticulture, livestock, fishery, agro-forestry, apiculture etc. to enable farmers not only in maximizing farm returns for sustaining livelihood, but also to mitigate the impacts of drought, flood or other extreme weather events with the income opportunity from allied activities during crop damage.

2 Sub-Mission on Agro forestry (SMAF)

Sub-Mission on Agro forestry under the framework of National Mission for Sustainable Agriculture (NMSA) has been launched during 2016-17 with an outlay of Rs.935 core for a period of 4 years (2016-17 to 2019-20). The aim of the Sub-Mission is to expand the tree coverage on farmland in complementary with agricultural crops. Liberal transit regulation rule for transport of timber is a precondition for availing the benefit of the Agro-forestry programme by the States. For the Year 2016-17, budget provision of Rs. 75.0 core has been made to implement the programme.

3 National Agro Forestry Policy

National Agro forestry Policy, 2014 has been formulated with the objective to bring coordination, convergence and synergy between various elements of agro forestry scattered in various existing Missions, programmes and schemes of the Government. The policy will provide a platform to stimulate the growth of agro forestry in India. It will encourage farmers for agro forestry as an integral component of farming system which will lead to meet the ever increasing demand of timber, food, fuel, fodder, fertilizer, fibber, and other agro forestry products; conserving the natural resources and forest; protecting the environment & providing environmental security; and increasing the forest / tree cover.

A major role for agro forestry is emerging in the domain of environmental services. Agro forestry is known to have the potential to mitigate the climate change effects through microclimate moderation and natural resources conservation in the short run and through carbon sequestration in the long run. Agro forestry species are known to sequester as much carbon in below ground biomass as the primary forests, and far greater than the crop and grass systems.

4 Pradhan Mantri Krishi Sinchayee Yojana (PMKSY)

The Pradhan Mantri Krishi Sinchayee Yojana (PMKSY) was launched on 1st July, 2015 with the motto of 'Her Khet Ko Paani' for providing end-to end solutions in irrigation supply chain, viz. water sources, distribution network and farm level applications. PMKSY not only focuses on creating sources for assured irrigation, but also creating protective irrigation by harnessing rain water at micro level through 'Jal Sanchay' and 'Jal Sinchan'. Micro irrigation is to be popularised to ensure 'Per drop-More crop'. PMKSY adopts State level planning and projectised execution that allows States to draw up their own irrigation development based on District Irrigation Plans and State Irrigation Plans.

PMKSY has the following components

- Accelerated Irrigation Benefit Programme(AIBP): To focus on faster completion of ongoing Major and Medium Irrigation, including National Projects. This component is being implemented by Ministry of Water Resources, RD &GR.
- PMKSY (Har Khet ko Pani): Source augmentation, distribution, ground water development, lift irrigation, diversion of water from water plenty to water scarce.areas, supplementing rain water harvesting beyond IWMP & MGNREGA, repair, restoration, renovation of traditional water bodies. This component is being implemented by Ministry of Water Resources, RD &GR.
- PMKSY (Watershed): Ridge area treatment, drainage line treatment, soil and moisture conservation, water harvesting structure, livelihood support activities and other watershed works. This component is being implemented by Department of Land Resources.
- PMKSY (Per Drop More Crop) being implemented by DAC & FW. This component includes two sub-components i.e. Micro irrigation and Other interventions. Per Drop More Crop component supports Micro level storage structures, efficient water conveyance & application, precision irrigation systems, topping up of input cost beyond MGNREGA permissible limits,

creation of secondary storage, water lifting devices etc. For the Year 2016-17, budget provision of Rs. 2340.0 crore has been made to implement the Per Drop More Crop component. The details of allocation, release and utilisation of Central Share reported by States under Micro irrigation and Other interventions sub components are as under:

PMKSY (Per Drop More Crop) - Micro Irrigation

Since January 2006, the Government of India has been implementing Centrally Sponsored Scheme on Micro Irrigation with the objective to enhance water use efficiency in the agriculture sector. Under the scheme, technological interventions like drip & sprinkler irrigation systems are promoted to encourage the farmers to use them for conservation and saving of water & improved yield.

In June 2010, the scheme was up-scaled to National Mission on Micro Irrigation (NMMI), which continued till the year 2013-14. From 1st April 2014, NMMI was subsumed under National Mission on Sustainable Agriculture (NMSA) and implemented as one of the components under "On Farm Water Management" (OFWM) during 2014-15.

From 1st April 2015, the scheme has been subsumed under Pradhan Mantri Krishi Sinchayee Yojana (PMKSY) and implemented as 'Per Drop More Crop (Micro Irrigation)' component of PMKSY.

Salient features of the micro irrigation scheme under PMKSY are as under:

- The rate of assistance (subsidy) comprising of both Central Govt. share and State Govt. share ranges from 35 to 60% for small, marginal and other farmers in DPAP/DDP & non-DPAP/DDP areas. The funding pattern between Central Govt. share and State Govt. share since November 2015 has been 60:40 for all states except North Eastern & Himalayan States. For North Eastern & Himalayan States the funding pattern is 90:10.
- Under Operational Guidelines of the scheme, indicative/average/ normative cost has been laid down for various crops/lateral spacing and different micro irrigation systems for the purpose of calculating financial assistance. Twenty five (25%) higher cost, over & above the normative cost for all systems has been fixed for North Eastern & Himalayan states for this purpose.
- Scheme is applicable in all states and all districts in the country. Support to each farmer is restricted to 5 ha per beneficiary. Assistance on the same land can be availed again only after a period of 10 years.
- Location and crop specific technologically appropriate irrigation systems are propagated under the scheme ensuring least cost burden to the farmers..

- All horticultural and agricultural crops are covered under the scheme. At least 25% of the funds allocated to the State under scheme are to be used necessarily for agricultural crops..
- At least 50% of the allocation to state is to be utilized for small and marginal farmers, of which, at least 30% has to be women beneficiaries/farmers. Further, 16% and 8% of the total allocation or in proportion of Scheduled Cast (SC)/Scheduled Tribe (ST) population is to be utilized for Special Component Sub-Plan (SCSP) and Tribal Sub-Plan (TSP) respectively..
- Equipments adhering to Indian Standards (BIS Standards) are only eligible for financial assistance under the scheme..
- Information & Communication Technology (ICT) has been extensively deployed for ensuring transparency in scheme implementation process and for effective monitoring of the programme. PMKSY website has been operationalized at the Central level. In many states also similar websites have been set up by the respective state governments.
- PMKSY has an end to end approach for development of irrigation chain from creation of water resources to field application of water in an integrated manner. Accordingly, all components of PMKSY are to be implemented in complementation with each other.

Adoption of micro irrigation is providing economic benefits to farmers :

Micro Irrigation which includes drip and sprinkler irrigation is a proven technology which has gained immense popularity amongst the farmers in India. Strengths of this technology include - efficient deployment of inputs such as water, electricity, fertilizers, labour etc, increase in crop productivity, better quality of produce leading to higher realization of sale price resulting in increased income of farmer and prosperity. With this technology, additional area can be irrigated with the same amount of water compared to conventional method of irrigation. In addition, water deficient, cultivable waste land and undulating land areas can easily be brought under cultivation due to ease of irrigation. With this technology, there is also a good scope for using it in closely spaced crops like rice, wheat, onion, potato etc. Benefits of this technology lead to control of ground and surface water pollution.

Micro irrigation technology has a good scope for generation of employment opportunities in rural areas besides arresting migration of farmers to cities.

Since implementation of Centrally Sponsored Scheme on Micro Irrigation in 2005-06, various studies have been undertaken to assess impact of this technology in various crops across different states and in the economic condition of the farmers adopting this technology. In these studies, the farmers have reported multiple benefits from adoption of micro irrigation technology. These benefits can be classified into two major categories - (i) the benefits that allow the farmer to cope up with adverse agricultural conditions (ii) benefits which bring prosperity to farmers and result in better economics as gross irrigated area increases & multi crop system enhances farmer's income resulting in financial gains to the farmers. Across these types, the benefits claimed by farmers are - saving in irrigation water due to improved irrigation efficiency resulting in reduced consumption of water, increase in crop yield, reduction in weeds, better growth of plants, reduction in consumption of various inputs like fertilizers, power, labour etc., better soil quality, increase in income of farmer and better returns to the farmers. The irrigated area also increases with adoption of micro irrigation systems besides change/shifting in cropping pattern from traditional crops to horticulture crops which are more remunerating and one can do timely irrigation of whole field in a single schedule.

The overall benefits accrued from the use of micro irrigation systems are reflected in substantial increase in income of farmers. The benefit cost (BC) ratio of installing micro irrigation systems is greater than "1" across all crops, which signifies importance of micro irrigation in enhancing net income of the farmers. The pay back period of micro irrigation technologies ranges from 1.5 to 2 years.

The positive outcomes of micro irrigation have made food security effective due to increase in production and productivity of different crops and increased area under irrigation from the same source of water resulting in enhanced nutritional security for the country.

Crop experiments have shown that use of water soluble fertilizers through drip irrigation (fertigation) can result in reduced consumption of fertilizers resulting in cost savings. Crop-wise details of increase in yields, savings in water & fertilizers at various locations is as under:

Government to launch new scheme to boost agriculture start-ups:

In an attempt to promote innovation and entrepreneurship in agriculture, the government is launching a new AGRI-UDAAN programme that will mentor startups and help them connect with potential investors.

Managed by India's premier farm research body, the Indian Council of Agricultural Research (ICAR), the food and agri-business accelerator programme to be launched on Friday will also help convert innovative ideas from India's rural youth into viable businesses.

"The idea is to attract the youth from rural India and elsewhere, and train them so they can add value to the farmers' produce," said Narendra Singh Rathore, deputy director general of agricultural education at ICAR.

Rathore added that following an intensive training lasting six months, the new start-ups will be connected to investors for funding.

"Our goal is to begin a start-up revolution in agriculture which so far has been limited to the services sector," he said.

Under the programme, start-ups will get incubation space to run their businesses and have access to research laboratories and libraries. AGRI UDAAN will also help the selected start-ups with regulatory services like company registration and environmental compliances.

In the pre-launch phase of the programme that began in 2015, the ICAR's National Academy of Agricultural Research Management (NARM) in Hyderabad incubated several new start-ups which are now processing exotic grains, manufacturing kinetic farm machines, and developing smart irrigation systems.

The companies include Hyderabad-based Inner Being, which processes locally grown and exotic millets and caters to the high-end wellness market. Another start-up, FlyBird Innovations in Bengaluru is developing smart irrigation techniques based on soil temperature, atmospheric humidity and water availability.

Kinemach, a company mentored under the UDAAN programme is developing a new range of farm machinery that mimics human tasks on the field-from ploughing to weeding.

5

MARKETING OF AGRICULTURAL PRODUCTS

Agricultural marketing comprises all the activities involved in the supply of farm inputs and output – including all those operations which are related to the procurement, collecting, grading, storing, food and agro-processing, transportation, financing and selling of the agricultural produce. In effect, marketing includes all overarching aspects of agribusiness, while it excludes the core activity of cultivation.

The agricultural marketing system also relates to economic growth of the agriculture sector and ensuring safe and affordable food to consumers, both of which are directly linked to the food security of the country.

Agricultural marketing is inferred to cover the services involved in moving an agricultural product from the farm to the consumer. It is also the planning, organizing, directing and handling of agricultural produce in such a way as to satisfy the farmer, producer and the consumer. Numerous interconnected activities are involved in doing this, such as planning production, growing and harvesting, grading, packing and packaging, transport, storage, agro- and food processing, distribution, advertisingand sale. Effectively, the term encompasses the entire range of supply chain operations. However, its key function is to help direct these services, by providing competent and able market information, thereby linking the other operations into an integrated service with targeted outcomes.

IMPORTANCE OF AGRICULTURAL AND FOOD MARKETING

In many countries, and virtually every less developed country (LDC), agriculture is the biggest single industry. Agriculture typically employs over fifty percent of the labour force in LDCs with industry and commerce dependent upon it as a source of raw materials and as a market for manufactured goods. Hence many argue that the development of agriculture and the marketing systems which impinge upon it are at the heart of the economic growth process in LDCs. Moreover as Kriesberg1 points out; in LDCs the consumer frequently spends in excess of fifty percent of the household's income on basic foodstuffs - much of which is inadequate both in quality and nutritional content.

By contrast Americans spend approximately twelve percent of their total disposable income on food. In Western Europe the figure ranges from about sixteen to nineteen percent of disposable income. Furthermore, whereas in developed countries the poor are relatively few in number, and therefore it is economically possible to establish special food distribution programmes to meet their needs, the scale of poverty in most LDCs is such that the commercial marketing system must be relied upon to perform the task of food distribution to poor and not-so-poor alike. This being so, it is imperative that the marketing system performs efficiently.

Economic development itself provides the impulse towards more sophisticated and more efficient marketing systems. Dixie2 suggests that as countries experience economic growth, their rate of urbanisation tends to increase substantially. Whereas the rate of population growth, in developing countries, averages around three percent per annum, their cities and towns are increasing their populations at about four percent per annum. In essence, this means that the number of people, in urban areas, needing to be fed by rural people, will double within sixteen years.

This has clear implications for agricultural production and the marketing systems that direct that production and distribute the output to the points of its consumption. Subsistence farming is likely to diminish in importance as farmers respond to the increased opportunities that development and urbanisation create; farms are likely to decrease in number whilst increasing in size; and agriculture will probably become less labour intensive and more capital intensive.

Dixie also highlights the potential contribution of agricultural and food marketing, towards attempts to improve rural incomes in developing countries. The inequality of incomes between the rural and urban areas draws people away from agricultural production and places great stress upon the infrastructure and social services of a country's towns and cities. Nowhere was this more dramatically demonstrated than in Nigeria when petroleum oil was discovered and then exploited in the 1970s. A large number of jobs were created in the urban areas and people abandoned agricultural production in large numbers. Nigeria became a net importer of many agricultural products of which it had formerly been a net exporter.

For as long as the world price for petroleum remained high the economy thrived and could well afford the food import bill. However, as soon as the world price for oil fell, the food import bill became a serious burden. Nigeria would only have avoided this scenario if it had been able to motivate people to continue in agriculture and this would only have been possible if the disparity between urban and rural incomes had been reduced. Rurally based enterprises, including small-holdings, can greatly improve their earning potential by adopting a market orientation. They can be encouraged to add value to commodities by adding to their utility. Value added products normally carry a higher margin than raw commodities.

Another development which has in recent times increased interest in marketing practies is the trend, in many developing countries, towards market liberalisation as part of economic structural adjustment programmes (ESAPs). The view that direct and indirect government participation in production and distribution had brought about structural distortions in economies has become widely accepted. Measures intended to correct these distortions include a return to market prices for all products and resources, the encouragement of a competitive private sector and the commercialisation, and sometimes privatisation, of all or some of the functions of marketing parastatals. All of this requires a better understanding of marketing practices and processes within the country implementing ESAPs, in general, and within the agricultural marketing parastatals affected, in particular.

So far this discussion has been set in the context of commercial marketing but social marketing should also be acknowledged. Social marketing identifies human needs in non-competitive economies and/or sectors of society and defines the means of delivering products and services to meet these needs. The marketing mix of social marketing strategies is evaluated using quite different criteria from those employed in assessing purely commercial marketing strategies. Criteria such as the percentage of the target population reached with the technology, products, processes or services, quantities produced and distributed and uptake of the product, service or technology are more often employed. Benefits are measured in terms of development goals, such as improved nutritional status or increased rural incomes. The use of economic criteria is usually limited to the latter and to selecting the least-cost strategy to achieve a quantitative goal. However, the criteria used to evaluate commercial marketing strategies should not automatically be eliminated, because these improve the efficiency of some aspects of social marketing strategy without preventing the attainment of social objectives.

The marketing concept and marketing systems

Marketing is not simply an extension of the production process but its only purpose as Adam Smith emphasised when, in his text The Wealth of Nations

(1776), he said that: "Consumption is the sole end purpose of all production: and the interest of the producer ought to be attended to only so far as it may be necessary for promoting that of the consumer."

Dixie relates what he describes as a definition of marketing which is:

"The series of services involved in moving a product (or commodity) from the point of production to the point of consumption."

This is a definition which many organisations, and governments, would recognise as describing their own activities in commodity marketing. Indeed in many developing countries it aptly describes, or in some cases, did in the past describe, the functions carried out by marketing parastatals with respect to staple foods. However, as Dixie himself points out, the definition omits two key elements of any definition of marketing production to effuse the marketing concept, i.e. a customer orientation and inbuilt sustainability. Gaedeke and Tootelian offer an alternative definition which overcomes the problems caused by these two omissions:

"... a management orientation focusing all the activities of the organization on satisfying customer needs and wants, thereby helping achieve the organization's long-range objectives."

This definition promotes a customer orientation and since the organisation's long-term objectives will include it's own continued existence it takes account of the need for sustainability. Moreover, this definition of the marketing concept does not preclude non-profit making organisations. Marketing is just as relevant to development projects, aid agencies, extension service organisations, and the like, as it is to commercial enterprises. Thus the marketing concept is that an organisation achieves its goals through the provision of customer satisfaction. Put another way, marketing is the integrative force that matches production to customer needs and satisfaction. Marketing is not an activity to which an organisation turns its attention at the end of the production phase of operations. Rather marketing needs to be directing production in accordance with clear signals from the marketplace as what is needed by customers.

The marketing concept must be adopted throughout not only the entire organisation, but the entire marketing system. A system is a complex of interrelated component parts or sub-systems which have a defined common goal. Thus, an agricultural and food marketing system comprises all of the functions, and agencies who perform those activities, that are necessary in order to profitably exploit opportunities in the marketplace. Each of the components, or sub-systems, are independent of one another but a change in any one of them impacts on the others as well as upon the system as a whole.

There is a danger that the marketing concept will be adopted by some parts of the system but not others. Thus, for example, a food manufacturer may be trying hard to implement the marketing concept and offer products that meet the precise needs of a target market. If, however, the manufacturer has to rely upon a farming community that is still very much production oriented, for raw material supplies, then the overall marketing objectives may be frustrated. In the same way, if only some functions are performed according to the marketing concept then the system as a whole may not achieve a market orientation. For instance, the marketing department may set out to serve the market for a high quality fruits and vegetables, for which it can obtain premium prices, but if transportation is performed using the same open-topped bulk carrying wagons used to ship grain and other aggregates then it is unlikely that the enterprise will deliver the product in the right condition for the target market.

AGRICULTURAL MARKETING DEVELOPMENT

Efforts to develop agricultural marketing have, particularly in developing countries, tended to concentrate on a number of areas, specifically infrastructure development; information provision; training of farmers and traders in marketing and post-harvest issues; and support to the development of an appropriate policy environment. In the past, efforts were made to develop government-run marketing bodies but these have tended to become less prominent over the years.

Agricultural Market infrastructure

Efficient marketing infrastructure such as wholesale, retail and assembly markets and storage facilities is essential for cost-effective marketing, to minimize post-harvest losses and to reduce health risks. Markets play an important role in rural development, income generation, food security, and developing rural-market linkages. Experience shows that planners need to be aware of how to design markets that meet a community's social and economic needs and how to choose a suitable site for a new market. In many cases sites are chosen that are inappropriate and result in under-use or even no use of the infrastructure constructed. It is also not sufficient just to build a market: attention needs to be paid to how that market will be managed, operated and maintained.

Rural assembly markets are located in production areas and primarily serve as places where farmers can meet with traders to sell their products. These may be occasional (perhaps weekly) markets, such as haat bazaars in India and Nepal, or permanent. Terminal wholesale markets are located in major metropolitan areas, where produce is finally channelled to consumers through trade between wholesalers and retailers, caterers, etc. The characteristics of wholesale markets have changed considerably as retailing changes in response to urban growth, the increasing role of supermarkets and increased consumer spending capacity. These changes may require responses in the way in which traditional wholesale markets are organized and managed.

Retail marketing systems in western countries have broadly evolved from traditional street markets through to the modern hypermarket or out-of-town shopping center. In developing countries, there remains scope to improve agricultural marketing by constructing new retail markets, despite the growth of supermarkets, although municipalities often view markets primarily as sources of revenue rather than infrastructure requiring development. Effective regulation of markets is essential. Inside a market, both hygiene rules and revenue collection activities have to be enforced. Of equal importance, however, is the maintenance of order outside the market. Licensed traders in a market will not be willing to cooperate in raising standards if they face competition from unlicensed operators outside who do not pay any of the costs involved in providing a proper service.

Market information

Efficient market information can be shown to have positive benefits for farmers and traders. Up-to-date information on prices and other market factors enables farmers to negotiate with traders and also facilitates spatial distribution of products from rural areas to towns and between markets. Most governments in developing countries have tried to provide market information services to farmers, but these have tended to experience problems of sustainability. Moreover, even when they function, the service provided is often insufficient to allow commercial decisions to be made because of time lags between data collection and dissemination. Modern communications technologies open up the possibility for market information services to improve information delivery through SMS on cell phones and the rapid growth of FM radio stations in many developing countries offers the possibility of more localised information services.

In the longer run, the internet may become an effective way of delivering information to farmers. However, problems associated with the cost and accuracy of data collection still remain to be addressed. Even when they have access to market information, farmers often require assistance in interpreting that information. For example, the market price quoted on the radio may refer to a wholesale selling price and farmers may have difficulty in translating this into a realistic price at their local assembly market. Various attempts have been made in developing countries to introduce commercial market information services but these have largely been targeted at traders, commercial farmers or exporters. It is not easy to see how small, poor farmers can generate sufficient income for a commercial service to be profitable although in India a new service introduced by Thomson Reuters was reportedly used by over 100,000 farmers in its first year of operation. Esoko in West Africa attempts to subsidize the cost of such services to farmers by charging access to a more advanced feature set of mobile-based tools to businesses.

Marketing training

Farmers frequently consider marketing as being their major problem. However, while they are able to identify such problems as poor prices, lack of transport and high post-harvest losses, they are often poorly equipped to identify potential solutions. Successful marketing requires learning new skills, new techniques and new ways of obtaining information. Extension officers working with ministries of agriculture or NGOs are often well-trained in agricultural production techniques but usually lack knowledge of marketing or post-harvest handling.

Enabling environments

Agricultural marketing needs to be conducted within a supportive policy, legal, institutional, macro-economic, infrastructural and bureaucratic environment. Traders and others are generally reluctant to make investments in an uncertain policy climate, such as those that restrict imports and exports or internal produce movement. Businesses have difficulty functioning when their trading activities are hampered by excessive bureaucracy. Inappropriate law can distort and reduce the efficiency of the market, increase the costs of doing business and retard the development of a competitive private sector. Poor support institutions, such as agricultural extension services, municipalities that operate markets inefficiently and inadequate export promotion bodies, can be particularly damaging. Poor roads increase the cost of doing business, reduce payments to farmers and increase prices to consumers. Finally, corruption can increase the transaction costs faced by those in the marketing chain.

AGRICULTURAL MARKETING SUPPORT

Most governments have at some stage made efforts to promote agricultural marketing improvements. In the United States the Agricultural Marketing Service (AMS) is a division of USDA and has programs that provide testing, support standardization and grading and offer market news services. AMS oversees marketing agreements and orders research and promotion programs. It also purchases commodities for federal food programs. USDA also provides support to agricultural marketing work at various universities. In the United Kingdom, support for marketing of some commodities was provided before and after the Second World War by boards such as the Milk Marketing Board and the Egg Marketing Board. These boards were closed down in the 1970s. As a colonial power, Britain established marketing boards in many countries, particularly in Africa. Some continue to exist although many were closed at the time of the introduction of structural adjustment measures in the 1990s.

Several developing countries have established government-sponsored marketing or agribusiness units. South Africa, for example, started the National Agricultural Marketing Council (NAMC) as a response to the deregulation of the agriculture industry and closure of marketing boards in the country. India has the long-established National Institute of Agricultural Marketing. These are primarily research and policy organizations, but other agencies provide facilitating services for marketing channels, such as the provision of infrastructure, market information and documentation support. Examples from the Caribbean include the National Agricultural Marketing Development Corporation (NAMDEVCO) in Trinidad and Tobago and the New Guyana Marketing Corporation in Guyana.

New marketing linkages between agribusiness, large retailers and farmers are gradually being developed, e.g. through contract farming, group marketing and other forms of collective action. Donors and NGOs are paying increasing attention to ways of promoting direct linkages between farmers and buyers within a value chain context. More attention is now being paid to the development of regional markets (e.g. East Africa) and to structured trading systems that should facilitate such developments. The growth of supermarkets, particularly in Latin America and East and South East Asia, is having a significant impact on marketing channels for horticultural, dairy and livestock products.Nevertheless, "spot" markets will continue to be important for many years, necessitating attention to infrastructure improvement such as for retail and wholesale markets.

Energy crop

An energy crop is a plant grown as a low-cost and low-maintenance harvest used to make biofuels, such as bioethanol, or combusted for its energy content to generate electricity or heat. Energy crops are generally categorized as woody or herbaceous plants; many of the latter are grasses of the family Graminaceae.

Commercial energy crops are typically densely planted, high-yielding crop species which are processed to bio-fuel and burnt to generate power. Woody crops such as willow or poplar are widely utilised, as well as temperate grasses such as Miscanthus and Pennisetum purpureum (both known as elephant grass). If carbohydrate content is desired for the production of biogas, whole-crops such as maize, Sudan grass, millet, white sweet clover, and many others can be made into silage and then converted into biogas.

Through genetic modification and application of biotechnology plants can be manipulated to create greater yields, high energy yields can also be realized with existing cultivars.. However, some additional advantages such as reduced associated costs (i.e. costs during the manufacturing process) and less water use can only be accomplished by using Genetically_modified_crops#Biofuel genetically modified crops.

Types

Energy generated by burning plants grown for the purpose, often after the dry matter is pelletized. Energy crops are used for firing power plants, either alone or co-fired with other fuels. Alternatively they may be used for heat or combined heat and power (CHP) production.

To cover the increasing requirements of woody biomass, short rotation coppice (SRC) were applied to agricultural sites. Within this cropping systems fast growing tree species like willows and poplars are planted in growing cycles of three to five years. The cultivation of this cultures is dependent on wet soil conditions and could be an alternative for moist field sieds. However, an influence on local water conditions could not be excluded. This indicates that an establishment should exclude the vicinity to vulnerable wetland ecosystems.

Gas biomass (methane)

Anaerobic digesters or biogas plants can be directly supplemented with energy crops once they have been ensiled into silage. The fastest growing sector of German biofarming has been in the area of "Renewable Energy Crops" on nearly 500,000 ha (1,200,000 acres) of land (2006). Energy crops can also be grown to boost gas yields where feedstocks have a low energy content, such as manures and spoiled grain. It is estimated that the energy yield presently of bioenergy crops converted via silage to methane is about 2 GWh/km (1.8×10 BTU/sq mi). Small mixed cropping enterprises with animals can use a portion of their acreage to grow and convert energy crops and sustain the entire farms energy requirements with about one fifth of the acreage. In Europe and especially Germany, however, this rapid growth has occurred only with substantial government support, as in the German bonus system for renewable energy. Similar developments of integrating crop farming and bioenergy production via silage-methane have been almost entirely overlooked in N. America, where political and structural issues and a huge continued push to centralize energy production has overshadowed positive developments.

Liquid biomass

European production of biodiesel from energy crops has grown steadily in the last decade, principally focused on rapeseed used for oil and energy. Production of oil/biodiesel from rape covers more than 12,000 km^2 in Germany alone, and has doubled in the past 15 years.Typical yield of oil as pure biodiesel may be is 100,000 L/km (68,000 US gal/sq mi; 57,000 imp gal/sq mi) or more, making biodiesel crops economically attractive, provided sustainable crop rotations exist that are nutrient-balanced and preventative of the spread of disease such as clubroot. Biodiesel yield of soybeans is significantly lower than that of rape.

Bioethanol

Energy crops for biobutanol are grasses. Two leading non-food crops for the production of cellulosic bioethanol are switchgrass and giant miscanthus. There has been a preoccupation with cellulosic bioethanol in America as the agricultural structure supporting biomethane is absent in many regions, with no credits or bonus system in place. Consequently, a lot of private money and investor hopes are being pinned on marketable and patentable innovations in enzyme hydrolysis and the like.

Bioethanol also refers to the technology of using principally corn (maize seed) to make ethanol directly through fermentation, a process that under certain field and process conditions can consume as much energy as is the energy value of the ethanol it produces, therefore being non-sustainable. New developments in converting grain stillage (referred to as distillers grain stillage or DGS) into biogas energy looks promising as a means to improve the poor energy ratio of this type of bioethanol process.

By dedication

Dedicated energy crops are non-food energy crops as giant miscanthus, switchgrass, jatropha, fungi, and algae. Dedicated energy crops are promising cellulose sources that can be sustainably produced in many regions of the United States. Additionally, the green waste byproducts of food and non-food energy crops can be used to produce various biofuels.

Agricultural value chain

The agricultural value chain concept has been used since the beginning of the millennium, primarily by those working in agricultural development in developing countries. Although there is no universally accepted definition of the term, it normally refers to the whole range of goods and services necessary for an agricultural product to move from the farm to the final customer or consumer.

The term value chain was first popularized in a book published in 1985 by Michael Porter, who used it to illustrate how companies could achieve what he called "competitive advantage" by adding value within their organization. Subsequently, the term was adopted for agricultural development purposes and has now become very much in vogue among those working in this field, with an increasing number of bilateral and multilateral aid organisations using it to guide their development interventions.

At the heart of the agricultural value chain concept is the idea of actors connected along a chain producing and delivering goods to consumers through a sequence of activities. However, this "vertical" chain cannot function in isolation and an important aspect of the value chain approach is that it also considers "horizontal" impacts on the chain, such as input and finance provision, extension support and the general enabling environment. The approach has been found useful, particularly by donors, in that it has resulted in a consideration of all those factors impacting on the ability of farmers to access markets profitably, leading to a broader range of chain interventions. It is used both for upgrading existing chains and for donors to identify market opportunities for small farmers.

Definitions

There is no commonly agreed definition of what is actually meant by agricultural value chains. Indeed, some agencies are using the term without having a workable definition or definitions and simply redefined ongoing activities as "value chain" work when the term came into vogue. Published definitions include the World Bank's "the term ''value chain" describes the full range of value adding activities required to bring a product or service through the different phases of production, including procurement of raw materials and other inputs", UNIDO's "actors connected along a chain producing, transforming and bringing goods and services to end-consumers through a sequenced set of activities", and CIAT's "a strategic network among a number of business organizations".

Without a universal definition the term "value chain" is now being used to refer to a range of types of chain, including:

- An international, or regional commodity market. Examples could include "the global cotton value chain", "the southern African maize value chain" or "the Brazilian coffee value chain";
- A national or local commodity market or marketing system such as "the Ghanaian tomato value chain" or ""the Accra tomato value chain";
- A supply chain, which can cover both of the above;

- An extended supply chain or marketing channel, which embraces all activities needed to produce the product, including information/extension, planning, input supply and finance. It is probably the most common usage of the value chain term;
- A dedicated chain designed to meet the needs of one or a limited number of buyers. This usage, which is arguably most faithful to Porter's concept, stresses that a value chain is designed to capture value for all actors by carrying out activities to meet the demand of consumers or of a particular retailer, processor or food service company supplying those consumers. Emphasis is firmly placed on demand as the source of the value.

Value chain methodologies

Donors and others supporting agricultural development, such as GIZ, DFID, ILO, IIED and UNIDO, have produced a range of documents designed to assist their staff and others to evaluate value chains in order to decide on the most appropriate interventions to either update existing chains or promote new ones. However, the application of value chain analysis is being interpreted differently by different organisations, with possible repercussions for their development impact. The proliferation of guides has taken place in an environment where key conceptual and methodological elements of value chain analysis and development are still evolving. Many of these guides include not only detailed procedures that require experts to carry out the analysis but also use detailed quasi-academic methodologies. One such methodology is to compare the same value chain over time (a comparative or panel study) to assess changes in rents, governance, systemic efficiency and the institutional framework.

LINKING FARMERS TO MARKETS

A major subset of value chain development work is concerned with ways of linking producers to companies, and hence into the value chains. While there are examples of fully integrated value chains that do not involve smallholders (e.g. Unilever operates tea estates and tea processing facilities in Kenya and then blends and packs the tea in Europe before selling it as Lipton, Brooke Bond or PG Tips brands), the great bulk of agricultural value chains involve sales to companies from independent farmers. Such arrangements frequently involve contract farming in which the farmer undertakes to supply agreed quantities of a crop or livestock product, based on the quality standards and delivery requirements of the purchaser, often at a price that is established in advance. Companies often also agree to support the farmer through input supply, land preparation, extension advice and transporting produce to their premises.

Inclusive value chains

Work to promote market linkages in developing countries is often based on the concept of "inclusive value chains", which usually places emphasis on identifying possible ways in which small-scale farmers can be incorporated into existing or new value chains or can extract greater value from the chain, either by increasing efficiency or by also carrying out activities further along the chain. In the various publications on the topic the definition of "inclusion" is often imprecise as it is often unclear whether the development aim is to include all farmers or only those best able to take advantage of the opportunities.

Agricultural value chain finance

Agricultural value chain finance is concerned with the flows of funds to and within a value chain to meet the needs of chain actors for finance, to secure sales, to buy inputs or produce, or to improve efficiency. Examining the potential for value chain finance involves a holistic approach to analyze the chain, those working in it, and their inter-linkages. These linkages allow financing to flow through the chain. For example, inputs can be provided to farmers and the cost can be repaid directly when the product is delivered, without need for farmers taking a loan from a bank or similar institution. This is common under contract farming arrangements. Types of value chain finance include product financing through trader and input supplier credit or credit supplied by a marketing company or a lead firm. Other trade finance instruments include receivables financing where the bank advances funds against an assignment of future receivables from the buyer, and factoring in which a business sells its accounts receivable at a discount. Also falling under value chain finance are asset collateralization, such as on the basis of warehouse receipts, and risk mitigation, such as forward contracting, futures and insurance.

The use of ICTs in value chains

Information and Communication Technologies, or ICTs, have become an important tool in promoting agricultural value chain efficiency. There has been a rapid expansion in the use of mobile technologies, in particular. The price of ICT services is falling and the technologies are becoming more affordable to many in developing countries. Applications can support farmers directly through SMS messages. Examples include iCow, developed in Kenya, which provides information on the gestation period, on artificial insemination of the cows, and on how to look after them. Applications such as M-Pesa can support access to mobile payment services for a large percentage of those without banks, thereby facilitating transactions in the value chain. Other applications have been developed to promote provision of crop insurance through input dealers, for example.

ICTs are also being used to strengthen the capacity of extension officers and NGO field staff to reach farmers with timely and accurate information and, at the same time, help capture data from the field. The Grameen Foundation's Community Knowledge Worker (CKW) programme is an example. Farmer representatives are trained to use ICT applications on a smartphone to provide agricultural information and extension support. Most market price information is now delivered to farmers via SMS. Further along the chain, technologies offer considerable possibilities to enhance traceability, which is particularly relevant as certification grows in importance. Where necessary many exporters can now trace consignments back to individual farmers and take necessary measures to address problems. Finally, systems such as FARA's eRails are also supporting agricultural researchers through data collection and analysis and access to up-to-date research publications.

Enabling environments

As with all agricultural growth, two things appear essential for successful value chain development: creating the right environment for agriculture and investing in rural public goods. An enabling environment implies peace and public order, macro-economic stability, inflation under control, exchange rates based on market fundamentals rather than government allocation of foreign currency, predictable taxation that is reinvested in public goods and property rights. There is a positive correlation of agricultural growth with investment in irrigation, transport infrastructure and other technologies. Governments have a responsibility to provide essential goods and services, infrastructure, such as rural roads, and agricultural research and extension. Value chain development is often constrained by corruption, both at a high level and at the ubiquitous road blocks found in many countries, particularly in Africa. Many measures to improve value chains require collaboration between a wide range of different ministries, and this can be difficult to achieve.

Intensive farming

Intensive farming involves various types of agriculture with higher levels of input and output per cubic unit of agricultural land area. It is characterized by a low fallow ratio, higher use of inputs such as capital and labour, and higher crop yields per cubic unit land area.This contrasts with traditional agriculture, in which the inputs per unit land are lower. The term "intensive" involves various meanings, some of which refer to organic farming methods (such as biointensive agriculture and French intensive gardening), and others that refer to nonorganic and industrial methods. Intensive animal farming involves either large numbers of animals raised on limited land, usually concentrated animal feeding operations (CAFOs), often referred to as factory farms, or managed intensive rotational grazing(MIRG), which has both organic and non-organic types. Both increase the yields of food and fiber per acre as compared to traditional animal husbandry. In CAFO, feed is brought to the seldom-moved animals, while in MIRG the animals are repeatedly moved to fresh forage.

Most commercial agriculture is intensive in one or more ways. Forms that rely heavily on industrial methods are often called industrial agriculture, which is characterised by innovations designed to increase yield. Techniques include planting multiple crops per year, reducing the frequency of fallow years, and improving cultivars. It also involves increased use of fertilizers, plant growth regulators, and pesticides and mechanised agriculture, controlled by increased and more detailed analysis of growing conditions, including weather, soil, water, weeds, and pests. This system is supported by ongoing innovation in agricultural machinery and farming methods, genetic technology, techniques for achieving economies of scale, logistics, and data collection and analysis technology. Intensive farms are widespread in developed nations and increasingly prevalent worldwide. Most of the meat, dairy, eggs, fruits, and vegetables available in supermarkets are produced by such farms.

Smaller intensive farms usually include higher inputs of labor and more often use sustainable intensive methods. The farming practices commonly found on such farms are referred to as appropriate technology. These farms are less widespread in both developed countries and worldwide, but are growing more rapidly. Most of the food available in specialty markets such as farmers markets is produced by these small holder farms.

Agricultural development in Britain between the 16th century and the mid-19th century saw a massive increase in agricultural productivity and net output. This in turn supported unprecedented population growth, freeing up a significant percentage of the workforce, and thereby helped enable the Industrial Revolution. Historians cited enclosure, mechanization, four-field crop rotation, and selective breeding as the most important innovations.

Industrial agriculture arose along with the Industrial Revolution. By the early 19th century, agricultural techniques, implements, seed stocks, and cultivars had so improved that yield per land unit was many times that seen in the Middle Ages.

The industrialization phase involved a continuing process of mechanization. Horse-drawn machinery such as the McCormick reaperrevolutionized harvesting, while inventions such as the cotton gin reduced the cost of processing. During this same period, farmers began to use steam-powered threshers and tractors, although they were expensive and dangerous. In 1892, the first gasoline-poweredtractor was successfully developed, and in 1923, the International Harvester Farmall tractor became the first all-purpose tractor, marking an inflection point in the replacement of draft animals with machines. Mechanical harvesters (combines), planters, transplanters, and other equipment were then developed, further revolutionizing agriculture. These inventions increased yields and allowed individual farmers to manage increasingly large farms.

The identification of nitrogen, phosphorus, and potassium (NPK) as critical factors in plant growth led to the manufacture of synthetic fertilizers, further increasing crop yields. In 1909, the Haber-Bosch method to synthesize ammonium nitrate was first demonstrated. NPK fertilizers stimulated the first concerns about industrial agriculture, due to concerns that they came with serious side effects such as soil compaction, soil erosion, and declines in overall soil fertility, along with health concerns about toxic chemicals entering the food supply.

The identification of carbon as a critical factor in plant growth and soil health, particularly in the form of humus, led to so-called sustainable agriculture, as well as alternative forms of intensive agriculture that also surpassed traditional agriculture, without side effects or health issues. Farmers adopting this approach were initially referred to as humus farmers, later as organic farmers.

The discovery of vitamins and their role in nutrition, in the first two decades of the 20th century, led to vitamin supplements, which in the 1920s allowed some livestock to be raised indoors, reducing their exposure to adverse natural elements. Chemicals developed for use in World War II gave rise to synthetic pesticides.

Following World War II, synthetic fertilizer use increased rapidly, while sustainable intensive farming advanced much more slowly. Most of the resources in developed nations went to improving industrial intensive farming, and very little went to improving organic farming. Thus, particularly in the developed nations, industrial intensive farming grew to become the dominant form of agriculture.

The discovery of antibiotics and vaccines facilitated raising livestock in confined animal feeding operations by reducing diseases caused by crowding. Developments in logistics and refrigeration as well as processing technology made long-distance distribution feasible.

Between 1700 and 1980, "the total area of cultivated land worldwide increased 466%" and yields increased dramatically, particularly because of selectively-bred, high-yielding varieties, fertilizers, pesticides, irrigation, and machinery. Global agricultural production doubled between 1820 and 1920; between 1920 and 1950; between 1950 and 1965; and again between 1965 and 1975 to feed a global population that grew from one billion in 1800 to 6.5 billion in 2002. The number of people involved in farming in industrial countries dropped, from 24 percent of the American population to 1.5 percent in 2002. In 1940, each farmworker supplied 11 consumers, whereas in 2002, each worker supplied 90 consumers. The number of farms also decreased and their ownership became more concentrated. In the year 2000 in the U.S., four companies produced 81 percent of cows, 73 percent of sheep, 57 percent of pigs, and 50 percent of chickens, which was cited as an example of "vertical integration" by the president of the U.S. National Farmers Union.Between 1967 and 2002, the one million pig farms in America consolidated into 114,000, with 80 million pigs (out of 95 million) produced each year on factory farms, according to the U.S. National Pork Producers Council. According to the Worldwatch Institute, 74 percent of the world's poultry, 43 percent of beef, and 68 percent of eggs are produced this way.

Concerns over the sustainability of industrial agriculture, which has become associated with decreased soil quality, and over the environmental effects of fertilizers and pesticides, have not subsided. Alternatives such as integrated pest management (IPM) have had little impact because policies encourage the use of pesticides and IPM is knowledge-intensive. These concerns sustained the organic movement and caused a resurgence in sustainable intensive farming, as well as funding for the development of appropriate technology.

Techniques and Technologies

Intensive livestock farming, also called "factory farming", is a term referring to the process of raising livestock in confinement at high stocking density. "Concentrated animal feeding operations" (CAFO), or "intensive livestock operations", can hold large numbers (some up to hundreds of thousands) of cows, hogs, turkeys, or chickens, often indoors. The essence of such farms is the concentration of livestock in a given space. The aim is to provide maximum output at the lowest possible cost and with the greatest level of food safety. The term is often used pejoratively. However, CAFOs have dramatically increased the production of food from animal husbandry worldwide, both in terms of total food produced and efficiency.

Food and water is delivered to the animals, and therapeutic use of antimicrobial agents, vitamin supplements, and growth hormones are often employed. Growth hormones are not used on chickens nor on any animal in the European Union. Undesirable behaviors often related to the stress of confinement led to a search for docile breeds (e.g., with natural dominant behaviors bred out), physical restraints to stop interaction, such as individual cages for chickens, or physical modification such as the de-beaking of chickens to reduce the harm of fighting.

The CAFO designation resulted from the 1972 U.S. Federal Clean Water Act, which was enacted to protect and restore lakes and rivers to a "fishable, swimmable" quality. The United States Environmental Protection Agency (EPA) identified certain animal feeding operations, along with many other types of industry, as "point source" groundwater polluters. These operations were subjected to regulation.

In 17 states in the U.S., isolated cases of groundwater contamination were linked to CAFOs. For example, the ten million hogs in North Carolina generate 19 million tons of waste per year. The U.S. federal government acknowledges the waste disposal issue and requires that animal waste be stored in lagoons. These lagoons can be as large as 7.5 acres (30,000 m). Lagoons not protected with an impermeable liner can leak into groundwater under some conditions, as can runoff from manure used as fertilizer. A lagoon that burst in 1995 released 25 million gallons of nitrous sludge in North Carolina's New River. The spill allegedly killed eight to ten million fish.

The large concentration of animals, animal waste, and dead animals in a small space poses ethical issues to some consumers. Animal rights and animal welfare activists have charged that intensive animal rearing is cruel to animals.

Other concerns include persistent noxious odor, the effects on human health, and the role of antibiotic use in the rise of resistant infectious bacteria.

According to the U.S. Centers for Disease Control and Prevention (CDC), farms on which animals are intensively reared can cause adverse health reactions in farm workers. Workers may develop acute and/or chronic lung disease, musculoskeletal injuries, and may catch (zoonotic) infections from the animals.

Managed intensive rotational grazing

Managed Intensive Rotational Grazing (MIRG), also known as cell grazing, mob grazing, and holistic management planned grazing, is a variety of foraging in which herds or flocks are regularly and systematically moved to fresh, rested grazing areas to maximize the quality and quantity of forage growth. MIRG can be used with cattle, sheep, goats, pigs, chickens, turkeys, ducks, and other animals. The herds graze one portion of pasture, or a paddock, while allowing the others to recover. Resting grazed lands allows the vegetation to renew energy reserves, rebuild shoot systems, and deepen root systems, resulting in long-term maximum biomass production. MIRG is especially effective because grazers thrive on the more tender younger plant stems. MIRG also leaves parasites behind to die off, minimizing or eliminating the need for de-wormers. Pasture systems alone can allow grazers to meet their energy requirements, and with the increased productivity of MIRG systems, the animals obtain the majority of their nutritional needs, in some cases all, without the supplemental feed sources that are required in continuous grazing systems.

Pasture intensification

Pasture intensification is the improvement of pasture soils and grasses to increase the food production potential of livestock systems. It is commonly used to reverse pasture degradation, a process characterized by loss of forage and decreased animal carrying capacity which results from overgrazing, poor nutrient management, and lack of soil conservation. This degradation leads to poor pasture soils with decreased fertility and water availability and increased rates of erosion, compaction, and acidification. Degraded pastures have significantly lower productivity and higher carbon footprints compared to intensified pastures.

Management practices which improve soil health and consequently grass productivity include irrigation, soil scarification, and the application of lime, fertilizers, and pesticides. Depending on the productivity goals of the target agricultural system, more involved restoration projects can be undertaken to replace invasive and under-productive grasses with grass species that are better suited to the soil and climate conditions of the region. These intensified grass systems allow higher stocking rates with faster animal weight gain and reduced time to slaughter, resulting in more productive, carbon-efficient livestock systems.

Another technique to optimize yield while maintaining the carbon balance is the use of integrated crop-livestock (ICL) and crop-livestock-forestry (ICLF) systems, which combine several ecosystems into one optimized agricultural framework. These synergies between these systems provide benefits to pastures through optimal plant usage, improved feedand fattening rates, increased soil fertility and quality, intensified nutrient cycling, integrated pest control, and improved biodiversity. The introduction of certain legume crops to pastures increases carbon accumulation and nitrogen fixation in soils, while their digestibility helps animal fattening and reduces methane emissions from enteric fermentation.ICLF systems yield beef cattle productivity up to ten times that of degraded pastures, additional crop production from maize, sorghum, and soybean harvests, and greatly reduced greenhouse gas balances due to forest carbon sequestration.

Crops

The Green Revolution transformed farming in many developing countries. It spread technologies that had already existed, but had not been widely used outside of industrialized nations. These technologies included "miracle seeds", pesticides, irrigation, and synthetic nitrogen fertilizer.

Seeds

In the 1970s, scientists created strains of maize, wheat, and rice that are generally referred to as high-yielding varieties (HYV). HYVs have an increased nitrogen-absorbing potential compared to other varieties. Since cereals that absorbed extra nitrogen would typically lodge (fall over) before harvest, semi-dwarfing genes were bred into their genomes. Norin 10 wheat, a variety developed by Orville Vogel from Japanese dwarf wheat varieties, was instrumental in developing wheat cultivars. IR8, the first widely implemented HYV rice to be developed by the International Rice Research Institute, was created through a cross between an Indonesian variety named "Peta" and a Chinese variety named "Dee Geo Woo Gen."

With the availability of molecular genetics in Arabidopsis and rice the mutant genes responsible (reduced height (rht), gibberellin insensitive (gai1) and slender rice (slr1)) have been cloned and identified as cellular signalling components of gibberellic acid, a phytohormone involved in regulating stem growth via its effect on cell division. Photosynthetic investment in the stem is reduced dramatically as the shorter plants are inherently more mechanically stable. Nutrients become redirected to grain production, amplifying in particular the yield effect of chemical fertilizers.

HYVs significantly outperform traditional varieties in the presence of adequate irrigation, pesticides, and fertilizers. In the absence of these inputs, traditional varieties may outperform HYVs. They were developed as F1 hybrids, meaning seeds need to be purchased every season to obtain maximum benefit, thus increasing costs.

Crop rotation

Crop rotation or crop sequencing is the practice of growing a series of dissimilar types of crops in the same space in sequential seasons for benefits such as avoiding pathogen and pest buildup that occurs when one species is continuously cropped. Crop rotation also seeks to balance the nutrient demands of various crops to avoid soil nutrient depletion. A traditional component of crop rotation is the replenishment of nitrogen through the use of legumes and green manure in sequence with cereals and other crops. Crop rotation can also improve soil structure and fertility by alternating deep-rooted and shallow-rooted plants. A related technique is to plant multi-species cover crops between commercial crops. This combines the advantages of intensive farming with continuous cover and polyculture.

Irrigation

Crop irrigation accounts for 70% of the world's fresh water use. Flood irrigation, the oldest and most common type, is typically unevenly distributed, as parts of a field may receive excess water in order to deliver sufficient quantities to other parts. Overhead irrigation, using center-pivot or lateral-moving sprinklers, gives a much more equal and controlled distribution pattern. Drip irrigation is the most expensive and least-used type, but delivers water to plant roots with minimal losses.

Water catchment management measures include recharge pits, which capture rainwater and runoff and use it to recharge groundwater supplies. This helps in the replenishment of groundwater wells and eventually reduces soil erosion. Dammed rivers creating reservoirs store water for irrigation and other uses over large areas. Smaller areas sometimes use irrigation ponds or groundwater.

Weed control

In agriculture, systematic weed management is usually required, often performed by machines such as cultivators or liquid herbicide sprayers. Herbicides kill specific targets while leaving the crop relatively unharmed. Some of these act by interfering with the growth of the weed and are often based on plant hormones. Weed control through herbicide is made more difficult when the weeds become resistant to the herbicide. Solutions include:

- Cover crops (especially those with allelopathic properties) that out-compete weeds or inhibit their regeneration
- Multiple herbicides, in combination or in rotation
- Strains genetically engineered for herbicide tolerance
- Locally adapted strains that tolerate or out-compete weeds
- Tilling
- Ground cover such as mulch or plastic
- Manual removal
- Mowing
- Grazing
- Burning

Terracing

In agriculture, a terrace is a leveled section of a hilly cultivated area, designed as a method of soil conservation to slow or prevent the rapid surface runoff of irrigation water. Often such land is formed into multiple terraces, giving a stepped appearance. The human landscapes of rice cultivation in terraces that follow the natural contours of the escarpments, like contour ploughing, are a classic feature of the island of Bali and the Banaue Rice Terraces in Banaue, Ifugao, Philippines. In Peru, the Inca made use of otherwise unusable slopes by building drystone walls to create terraces.

Rice paddies

A paddy field is a flooded parcel of arable land used for growing rice and other semiaquatic crops. Paddy fields are a typical feature of rice-growing countries of east and southeast Asia, including Malaysia, China, Sri Lanka, Myanmar, Thailand, Korea, Japan, Vietnam, Taiwan, Indonesia, India, and the Philippines. They are also found in other rice-growing regions such as Piedmont (Italy), the Camargue (France), and the Artibonite Valley (Haiti). They can occur naturally along rivers or marshes, or can be constructed, even on hillsides. They require large water quantities for irrigation, much of it from flooding. It gives an environment favourable to the strain of rice being grown, and is hostile to many species of weeds. As the only draft animal species which is comfortable in wetlands, the water buffalo is in widespread use in Asian rice paddies.

Paddy-based rice-farming has been practiced in Korea since ancient times. A pit-house at the Daecheon-ni archaeological site yielded carbonized rice grains and radiocarbon dates indicating that rice cultivation may have begun as early as the Middle Jeulmun Pottery Period (c. 3500-2000 BC) in the Korean Peninsula. The earliest rice cultivation there may have used dry-fields instead of paddies.

The earliest Mumun features were usually in naturally swampy, low-lying narrow gulleys and fed by local streams. Some Mumun paddies in flat areas were made of a series of squares and rectangles separated by bunds approximately 10 cm in height, while terraced paddies were long and irregular in shape, following the natural contours of the land at various levels.

Like today, Mumun period rice farmers used terracing, bunds, canals, and small reservoirs. Some paddy-farming techniques of the Middle Mumun period (c. 850-550 BC) can be interpreted from the well-preserved wooden tools excavated from archaeological rice paddies at the Majeon-ni site. Iron tools for paddy-farming were not introduced until sometime after 200 BC. The spatial scale of individual paddies, and thus entire paddy-fields, increased with the regular use of iron tools in the Three Kingdoms of Korea Period (c.AD 300/400-668). A recent development in the intensive production of rice is System of Rice Intensification (SRI). Developed in 1983 by the French Jesuit Father Henri de Laulanié in Madagascar,by 2013 the number of smallholder farmers using SRI had grown to between 4 and 5 million.

Sustainable intensive farming

Sustainable intensive farming practices have been developed o slow the deterioration of agricultural land and even regenerate soil health and ecosystem services, while still offering high yields. Most of these developments fall in the category of organic farming, or the integration of organic and conventional agriculture. "Organic systems and the practices that make them effective are being picked up more and more by conventional agriculture and will become the foundation for future farming systems. They won't be called organic, because they'll still use some chemicals and still use some fertilizers, but they'll function much more like today's organic systems than today's conventional systems."

The System of Crop Intensification (SCI) was born out of research primarily at Cornell University and smallholder farms in India on SRI. It uses the SRI concepts and methods for rice and applies them to crops like wheat, sugarcane, finger millet, and others. It can be 100% organic, or integrated with reduced conventional inputs.

Holistic management is a systems thinking approach that was originally developed for reversing desertification. Holistic planned grazing is similar to rotational grazing but differs in that it more explicitly provides a framework for adapting to four basic ecosystem processes: the water cycle, the mineral cycle (including the carbon cycle), energy flow, and community dynamics (the relationship between organisms in an ecosystem) as equal in importance to livestock production and social welfare. By intensively managing the behavior and movement of livestock, holistic planned grazing simultaneously increases stocking rates and restores grazing land.

Pasture cropping involves planting grain crops directly into grassland without first applying herbicides. The perennial grasses form a living mulch understory to the grain crop, eliminating the need to plant cover crops after harvest. The pasture is intensively grazed both before and after grain production using holistic planned grazing. This intensive system yields equivalent farmer profits (partly from increased livestock forage) while building new topsoil and sequestering up to 33 tons of CO2/ha/year.

The Twelve Aprils grazing program for dairy production, developed in partnership with the USDA-SARE, is similar to pasture cropping, but the crops planted into the perennial pasture are forage crops for dairy herds. This system improves milk production and is more sustainable than confinement dairy production.

Integrated multi-trophic aquaculture (IMTA) is an example of a holistic approach. IMTA is a practice in which the by-products (wastes) from one species are recycled to become inputs (fertilizers, food) for another. Fed aquaculture (e.g. fish, shrimp) is combined with inorganic extractive (e.g. seaweed) and organic extractive aquaculture (e.g. shellfish) to create balanced systems for environmental sustainability (biomitigation), economic stability (product diversification and risk reduction), and social acceptability (better management practices).

Biointensive agriculture focuses on maximizing efficiency such as per unit area, energy input, and water input.

Agroforestry combines agriculture and orchard/forestry technologies to create more integrated, diverse, productive, profitable, healthy, and sustainable land-use systems.

Intercropping can increase yields or reduce inputs and thus represents (potentially sustainable) agricultural intensification. However, while total yield per acre is often increased dramatically, yields of any single crop often diminish. There are also challenges to farmers relying on farming equipment optimized for monoculture, often resulting in increased labor inputs.

Vertical farming is intensive crop production on a large scale in urban centers, in multi-story, artificially-lit structures, using far less inputs and producing fewer environmental impacts.

An integrated farming system is a progressive, biologically-integrated sustainable agriculture system such as IMTA or Zero waste agriculture, whose implementation requires exacting knowledge of the interactions of multiple species and whose benefits include sustainability and increased profitability. Elements of this integration can include:

- Intentionally introducing flowering plants into agricultural ecosystems to increase pollen-and nectar-resources required by natural enemies of insect pests
- Using crop rotation and cover crops to suppress nematodes in potatoes

Challenges

The challenges and issues of industrial agriculture for society, for the industrial agriculture sector, for the individual farm, and for animal rights include the costs and benefits of both current practices and proposed changes to those practices.

This is a continuation of thousands of years of invention in feeding ever-growing populations.

[W]hen hunter-gatherers with growing populations depleted the stocks of game and wild foods across the Near East, they were forced to introduce agriculture. But agriculture brought much longer hours of work and a less rich diet than hunter-gatherers enjoyed. Further population growth among shifting slash-and-burn farmers led to shorter fallow periods, falling yields and soil erosion. Plowing and fertilizers were introduced to deal with these problems - but once again involved longer hours of work and degradation of soil resources (Boserup, The Conditions of Agricultural Growth, Allen and Unwin, 1965, expanded and updated in Population and Technology, Blackwell, 1980.).

While the point of industrial agriculture is to profitably supply the world at the lowest cost, industrial methods have significant side effects. Further, industrial agriculture is not an indivisible whole, but instead is composed of multiple elements, each of which can be modified in response to market conditions, government regulation, and further innovation, and has its own side-effects. Various interest groups reach different conclusions on the subject.

Population growth

- 30,000 years ago hunter-gatherer behavior fed 6 million people
- 3,000 years ago primitive agriculture fed 60 million people
- 300 years ago intensive agriculture fed 600 million people
- Today industrial agriculture attempts to feed 8 billion people

Between 1930 and 2000, U.S. agricultural productivity (output divided by all inputs) rose by an average of about 2 percent annually, causing food prices to decrease. "The percentage of U.S. disposable income spent on food prepared at home decreased, from 22 percent as late as 1950 to 7 percent by the end of the century."

Other impacts

Industrial agriculture uses huge amounts of water, energy, and industrial chemicals, increasing pollution in the arable land, usable water, and atmosphere. Herbicides, insecticides, and fertilizers are accumulating in ground and surface waters. "Many of the negative effects of industrial agriculture are remote from fields and farms. Nitrogen compounds from the Midwest, for example, travel down the Mississippi to degrade coastal fisheries in the Gulf of Mexico. But other adverse effects are showing up within agricultural production systems-for example, the rapidly developing resistance among pests is rendering our arsenal of herbicides and insecticides increasingly ineffective."Agrochemicals and monoculture have been implicated in Colony Collapse Disorder, in which the individual members of bee colonies disappear. Agricultural production is highly dependent on bees to pollinate many varieties of fruits and vegetables.

Land banking

Land banking is the practice of aggregating parcels of land for future sale or development. While in many countries land banking may refer to various private real-estate investment schemes, in the United States it refers to the establishment of quasi-governmental county or municipal authorities purposed with managing an inventory of surplus land.

AGRICULTURAL MARKETING IN INDIA

India is an agricultural country and one third population depends on the agricultural sector directly or indirectly. Agriculture remains as the main stray of the Indian economy since times immemorial. Indian agriculture contribution to the national gross domestic product (GDP) is about 25 per cent. With food being the crowning need of mankind, much emphasis has been on commercialising agricultural production. For this reason, adequate production and even distribution of food has of late become a high priority global concern.

Agricultural marketing is mainly the buying and selling of agricultural products. In earlier days when the village economy was more or less self-sufficient the marketing of agricultural products presented no difficulty as the farmer sold his produce to the consumer on a cash or barter basis. Agricultural Marketing, rather than production, is going to be the key driver of the agriculture sector today, thanks to the new market realities posed by the increasing accent on globalization, liberalization and privatization of the economy. Market-driven production is an idea whose time has come. With the gradual shifting of agricultural system from subsistence to commercial one, there is increasing focus on Agripreneurship and Agri-marketing. It is the need of the time to tune up the Agricultural Marketing System of the country to enable the farmers to face the new challenges and reap the opportunities as well. This summons us to revisit our traditional statistic policies and laws and bring about the requisite reforms in the sector.

The imperatives of the integration of the Agricultural Marketing System of the country are increasingly posing new challenges in respect of each of the components of the system such as- cleaning, grading, quality certification, packaging, storage, transportation, financing, wholesaling and retailing etc of agriculture produce.

The subtlety of the management of Agricultural Marketing issues today calls for high degree of professionalism to measure up to the expectations of different stakeholders.

Internal reforms in traditional agricultural marketing system of the country have become indispensable to enable our farmers to tap into the external market of the post-WTO regime. Now that many states have introduced some reforms in agricultural marketing, the need of the hour is to consolidate the gain of reforms through appropriate policies and plans. The existing trade barriers have to be removed for evolving a common market in India. There is a need to give a push to processing and value addition. Cost effectiveness through integrated supply chain management has become sine qua non in the present world trade environment. This demands doing away with the dysfunctional aspects of the present regulated marketing system by putting in place a responsive market information system, need-based price discovery mechanism, measures to solve the problem of economies of scale at the grass root level by promoting contract farming, direct marketing and private investment in agricultural marketing infrastructure.

Congenial investor-friendly economic environment has got to be evolved to attract the private investors to invest in agricultural marketing sector for bridging the existing gap in the agricultural marketing infrastructure of the country. Farmers should get freedom of choice to use a physical market giving them services for the fees paid by them. This could be possible through dismantling of the monopoly of the Government-owned markets and permitting setting up markets in the private sector, thereby bringing healthy competition in the sector. This also calls for the traditional Government mandies to be spruced up in terms of their management, infrastructure, service delivery, customer-friendliness through fine-tuning of their internal processes. The other areas of reforms such as promotion of grading and standardization, market-led extension, rationalization of market fees, public-private partnership in owning and management of markets etc will go a long way towards pushing the system to the next level of excellence. The focus on reforms calls for a paradigm shift in our approach.

Today's agricultural marketing has to undergo a series of exchanges or transfers from one person to another before it reaches the consumer. There are three marketing functions involved in this, i.e., assembling, preparation for consumption and distribution. Selling on any agricultural produce depends on some couple of factors like the demand of the product at that time, availability of storage etc. The products may be sold directly in the market or it may be stored locally for the time being. Moreover, it may be sold as it is gathered from the field or it may be cleaned, graded and processed by the farmer or the merchant of the village. Sometime processing is done because consumers want it, or sometimes to conserve the quality of that product. The task of distribution system is to match the supply with the existing demand by whole selling and retailing in various points of different markets like primary, secondary or terminal markets.

Most of the agricultural products in India are sold by farmers in the private sector to moneylenders (to whom the farmer may be indebted) or to village traders. Products are sold in various ways. For example, it might be sold at a weekly village market in the farmer's village or in a neighboring village. If these outlets are not available, then produce might be sold at irregularly held markets in a nearby village or town, or in the mandi.

In India, there are several central government organisations, who are involved in agricultural marketing like, Commission of Agricultural Costs and Prices, Food Corporation of India, Cotton Corporation of India, Jute Corporation of India, etc. There are also specialised marketing bodies for rubber, tea, coffee, tobacco, spices and vegetables.

Under the Agricultural Produce (grading and marketing) Act of 1937, more than forty primary commodities are compulsorily graded for export and voluntarily graded for internal consumption. Although the regulation of commodity markets is a function of state government, the directorate of marketing and inspection provides marketing and inspection services and financial aid down to the village level to help set up commodity grading centers in selected markets.

As we have a tradition of agricultural production, marketing and allied commercial activities, now it is the time for us to brainstorm and come out with new ideas of value added services. These value added services will give the existing agricultural engine a new dimension. The next logical step could be food-processing which not only could be another revenue generating area but also can provide lots of full-time employment to our youths. With the changing agricultural scenario and global competition, there is a need of exploiting the available resources at maximum level.

There was a survey undertaken by the directorate of marketing and inspection in the ministry of agriculture in 1970-71 and 1971-72, of five hundred regulated markets was, with a view to assessing the adequacy and efficiency of the existing regulated markets and highlighting their drawbacks and deficiencies and suggesting measures to develop them. One of the most important drawbacks has been the inadequate financial resources of some of the market committees. During the fourth plan, a central sector scheme was drawn up by the ministry of agriculture to provide a grant at 20 per cent of the cost of development of market, subject to a maximum of Rs. 2 lakhs. The balance will have to be provided by the commercial banks.

Another important development in the field of regulated markets is the keen interest taken by the International Development Agency (IDA) in the development of the infrastructure in regulated markets. The IDA is financing the development of infrastructure in 50 markets of Bihar.

There are also some good news on the front of irrigation, rural infrastructure, restoring water bodies and water harvesting. Another action initiated to improve the governance of the Small Farmers Agri-business Consortium (SFAC) including the appointment of a banker as the chief executive; necessary additional capital to be provided to SFAC.

Agricultural Marketing is an area for the "second generation" of green revolution problems. Indian Marketing is undergoing a significant metamorphosis because of economic liberalization and globalisation. Infrastructure development is the critical factor for determining the success of market-oriented strategy and macro-economic policies in developing countries. Both national and global players are trying to capture the urban markets of India, which has already reached a saturation level, and it is extremely difficult to tap the urban markets with high profit margin. In export market, a firm has to face challenges from Multi-National Corporates (MNCs), other foreign companies and domestic firms of the concerned export markets. While urban market or export market is difficult to plough, rural market is relatively easy and feasible to cultivate, in India.

Market information is an important aspect of Agricultural Marketing. The importance of sound agricultural marketing policies for ensuring fair returns to the farmers can hardly be over-emphasized. It, therefore, becomes necessary on the part of regulatory agencies to ensure remunerative prices to the farmers for the sale of their produce, to boost up their efforts for increasing and sustaining the agricultural production. A number of measures have been taken by the Government to protect and safeguard the interests of farmers, like regulation of markets, grading of agricultural produce, cooperative marketing etc. Still the benefits are not percolating down to the farmers, as they are unable to plan their strategies for sale of their produce at remunerative prices, in the absence of correct and timely market information and advice about arrivals, prices, market trend, etc.

Almost all the States and Union Territories are providing market information in one form or the other for the benefits of market users like producers, traders and consumers. However, the information is collected and disseminated by use of conventional methods which cause inordinate delay in communicating the information to different target groups, and thus adversely affecting their economic interest. Therefore, there is an urgent need to bring improvement in the present market information system by linking (i) all Agricultural Produce Wholesale markets in the States and Union Territories, and (ii) the State Agricultural Marketing Boards and State Directorates of Agricultural Marketing, with the Directorate of Marketing & Inspection of the Ministry of Agriculture, for effective and efficient information exchange. In view of this importance, the Department of Agriculture & Cooperation, in the Union Ministry of Agriculture has sanctioned a Central Sector Scheme "NICNET Based Agricultural Marketing Information Network (AGMARK-NET)", Project Cost estimate of about Rs.10.00 Crores, to link 810 nodes (735 agricultural markets and 75 State Marketing Boards/DMI offices) to begin with, through its attached office, Directorate of Marketing and Inspection (DMI), during the year 2000-02. During the X five year plan, further expansion to 2000 additional markets, Regional Portals and GIS based National atlas of Markets has also been approved. The Scheme is being implemented during the XI Plan period by enhancing its scope to cover 500 additional markets including remaining principal regulated markets, if any, farmer/consumer markets, private markets, panchayat markets, urban markets and special commodity markets.

Introduction to Agricultural Marketing

Agricultural marketing covers the services involved in moving an agricultural product from the farm to the consumer. Numerous interconnected activities are involved in doing this, such as planning production, growing and harvesting, grading, packing, transport, storage, agro- and food processing, distribution, advertising and sale. Some definitions would even include "the acts of buying supplies, renting equipment, (and) paying labor", arguing that marketing is everything a business does. Such activities cannot take place without the exchange of information and are often heavily dependent on the availability of suitable finance.

Marketing systems are dynamic; they are competitive and involve continuous change and improvement. Businesses that have lower costs, are more efficient, and can deliver quality products, are those that prosper. Those that have high costs, fail to adapt to changes in market demand and provide poorer quality are often forced out of business. Marketing has to be customer-oriented and has to provide the farmer, transporter, trader, processor, etc. with a profit. This requires those involved in marketing chains to understand buyer requirements, both in terms of product and business conditions.

In Western countries considerable agricultural marketing support to farmers is often provided. In the USA, for example, the USDA operates the Agricultural

Marketing Service. Support to developing countries with agricultural marketing development is carried out by various donor organizations and there is a trend for countries to develop their own Agricultural Marketing or Agribusiness units, often attached to ministries of agriculture. Activities include market information development, marketing extension, training in marketing and infrastructure development. Since the 1990s trends have seen the growing importance of supermarkets and a growing interest in contract farming, both of which impact significantly on the way in which marketing takes place.

Agricultural marketing support

In the United States the Agricultural Marketing Service (AMS) is a division of USDA and has programs for cotton, dairy, fruit and vegetable, livestock and seed, poultry, and tobacco. These programs provide testing, standardization, grading and market news services and oversee marketing agreements and orders, administer research and promotion programs, and purchase commodities for federal food programs. The AMS also enforces certain federal laws. USDA also provides support to the Agricultural Marketing Resource Center at Iowa State University and toPenn State University.

In the United Kingdom support for marketing of some commodities was provided before and after the Second World War by boards such as the Milk Marketing Board and the Egg Marketing Board, but these were closed down in the 1970s. As a colonial power Britain established marketing boards in many countries, particularly in Africa. Some continue to exist although many were closed down at the time of the introduction of structural adjustment measures in the 1990s.

In recent years several developing countries have established government-sponsored marketing or agribusiness units. South Africa, for example, started the National Agricultural Marketing Council (NAMC) as a response to the deregulation of the agriculture industry and closure of marketing boards in the country. India has the long-established National Institute of Agricultural Marketing (NIAM). These are primarily research and policy organizations, but other agencies provide facilitating services for marketing channels, such as the provision of infrastructure, market information and documentation support. Examples include the National Agricultural Marketing Development Corporation (NAMDEVCO) in Trinidad and Tobago and the New Guyana Marketing Corporation.

Several organizations provide support to developing countries to develop their agricultural marketing systems, including FAO's agricultural marketing unit and various donor organizations. There has also recently been considerable interest by NGOs to carry out activities to link farmers to markets.

Agricultural marketing development

Well-functioning marketing systems necessitates a strong private sector backed up by appropriate policy and legislative frameworks and effective government support services. Such services can include provision of market infrastructure, supply of market information (as done by USDA, for example), and agricultural extension services able to advise farmers on marketing. Training in marketing at all levels is also needed. One of many problems faced in agricultural marketing in developing countries is the latent hostility to the private sector and the lack of understanding of the role of the intermediary. For this reason "middleman" has become very much a pejorative word.

Agricultural advisory services and the market

Promoting market orientation in agricultural advisory services aims to provide for the sustainable enhancement of the capabilities of the rural poor to enable them to benefit from agricultural markets and help them to adapt to factors which impact upon these. As a study by the Overseas Development Institute demonstrates, a value chain approach to advisory services indicates that the range of clients serviced should go beyond farmers to include input providers, producers, producer organisations and processors and traders.

Market infrastructure

Efficient marketing infrastructure such as wholesale, retail and assembly markets and storage facilities is essential for cost-effective marketing, to minimise post-harvest losses and to reduce health risks. Markets play an important role in rural development, income generation, food security, developing rural-market linkages and gender issues. Planners need to be aware of how to design markets that meet a community's social and economic needs and how to choose a suitable site for a new market. In many cases sites are chosen that are inappropriate and result in under-use or even no use of the infrastructure constructed. It is also not sufficient just to build a market: attention needs to be paid to how that market will be managed, operated and maintained. In most cases, where market improvements were only aimed at infrastructure upgrading and did not guarantee maintenance and management, most failed within a few years.

Rural assembly markets are located in production areas and primarily serve as places where farmers can meet with traders to sell their products. These may be occasional (perhaps weekly) markets, such as haat bazaars in India and Nepal, or permanent. Terminal wholesale markets are located in major metropolitan areas, where produce is finally channelled to consumers through trade between wholesalers and retailers, caterers, etc. The characteristics of wholesale markets have changed considerably as retailing changes in response to urban growth, the increasing role of supermarkets and increased consumer spending capacity. These changes require responses in the way in which traditional wholesale markets are organized and managed.

Retail marketing systems in western countries have broadly evolved from traditional street markets through to the modern hypermarket or out-of-town shopping centre. In developing countries, there remains considerable scope to improve agricultural marketing by constructing new retail markets, despite the growth of supermarkets, although municipalities often view markets as sources of revenue rather than infrastructure requiring development. Effective regulation of markets is essential. Inside the market, both hygiene rules and revenue collection activities have to be enforced. Of equal importance, however, is the maintenance of order outside the market. Licensed traders in a market will not be willing to cooperate in raising standards if they face competition from unlicensed operators outside who do not pay any of the costs involved in providing a proper service.

Market information

Efficient market information can be shown to have positive benefits for farmers and traders. Up-to-date information on prices and other market factors enables farmers to negotiate with traders and also facilitates spatial distribution of products from rural areas to towns and between markets. Most governments in developing countries have tried to provide market information services to farmers, but these have tended to experience problems of sustainability. Moreover, even when they function, the service provided is often insufficient to allow commercial decisions to be made because of time lags between data collection and dissemination. Modern communications technologies open up the possibility for market information services to improve information delivery through SMS on cell phones and the rapid growth of FM radio stations in many developing countries offers the possibility of more localised information services. In the longer run, the internet may become an effective way of delivering information to farmers.

However, problems associated with the cost and accuracy of data collection still remain to be addressed. Even when they have access to market information, farmers often require assistance in interpreting that information. For example, the market price quoted on the radio may refer to a wholesale selling price and farmers may have difficulty in translating this into a realistic price at their local assembly market. Various attempts have been made in developing countries to introduce commercial market information services but these have largely been targeted at traders, commercial farmers or exporters. It is not easy to see how small, poor farmers can generate sufficient income for a commercial service to be profitable although in India a new service introduced by Thompson Reuters was reportedly used by over 100,000 farmers in its first year of operation. Esoko in West Africa attempts to subsidize the cost of such services to farmers by charging access to a more advanced feature set of mobile-based tools to businesses.

Marketing training

Farmers frequently consider marketing as being their major problem. However, while they are able to identify such problems as poor prices, lack of transport and high post-harvest losses, they are often poorly equipped to identify potential solutions. Successful marketing requires learning new skills, new techniques and new ways of obtaining information. Extension officers working with ministries of agriculture or NGOs are often well-trained in horticultural production techniques but usually lack knowledge of marketing or post-harvest handling. Ways of helping them develop their knowledge of these areas, in order to be better able to advise farmers about market-oriented horticulture, need to be explored. While there is a range of generic guides and other training materials available from FAO and others, these should ideally be tailored to national circumstances to have maximum effect.

Enabling environments

Agricultural marketing needs to be conducted within a supportive policy, legal, institutional, macro-economic, infrastructural and bureaucratic environment. Traders and others cannot make investments in a climate of arbitrary government policy changes, such as those that restrict imports and exports or internal produce movement. Those in business cannot function if their trading activities are hampered by excessive bureaucracy. Inappropriate law can distort and reduce the efficiency of the market, increase the costs of doing business and retard the development of a competitive private sector. Poor support institutions, such as agricultural extension services, municipalities that operate markets inefficiently and export promotion bodies, can be particularly damaging. Poor roads increase the cost of doing business, reduce payments to farmers and increase prices to consumers. Finally, the ever-present problem of corruption can seriously impact on agricultural marketing efficiency in many countries by increasing the transaction costs faced by those in the marketing chain.

Recent developments

New marketing linkages between agribusiness, large retailers and farmers are gradually being developed, e.g. through contract farming, group marketing and other forms of collective action.Donors and NGOs are paying increasing attention to ways of promoting direct linkages between farmers and buyers. The growth of supermarkets, particularly in Latin America and East and South East Asia, is having a significant impact on marketing channels for horticultural, dairy and livestock products. Nevertheless, "spot" markets will continue to be important for many years, necessitating attention to infrastructure improvement such as for retail and wholesale markets.

Importance and Objectives of Agriculture Marketing

The farmer has realized the importance of adopting new techniques of production and is making efforts for more income and higher standards of living. As a consequence, the cropping pattern is no longer dictated by what he needs for his own personal consumption but what is responsive to the market in terms of prices received by him. While the trade is very organised the farmers are not Farmer is not conversant with the complexities of the marketing system which is becoming more and more complicated. The cultivator is handicapped by several disabilities as a seller. He sells his produce at an unfavorable place, time and price.

The objectives of an efficient marketing system are:

1. to enable the primary producers to get the best possible returns,
2. to provide facilities for lifting all produce, the farmers are willing, to sell at an incentive price,
3. to reduce the price difference between the primary producer and ultimate consumer, and
4. to make available all products of farm origin to consumers at reasonable price without impairing on the quality of the produce.

Facilities Needed for Agricultural Marketing

In order to have best advantage in marketing of his agricultural produce the farmer should enjoy certain basic facilities.

1. He should have proper facilities for storing his goods.
2. He should have holding capacity, in the sense, that he should be able to wait for times when he could get better prices for his produce and not dispose of his stocks immediately after the harvest when the prices are very low.
3. He should have adequate and cheap transport facilities which could enable him to take his surplus produce to the mandi rather than dispose it of in the village itself to the village money-lender-cum-merchant at low prices.

4. He should have clear information regarding the market conditions as well as about the ruling prices, otherwise may be cheated. There should be organized and regulated markets where the farmer will not be cheated by the "dalals" and "arhatiyas".
5. The number of intermediaries should be as small as possible, so that the middleman's profits are reduced. This increases! the returns to the farmers.

INADEQUACIES OF PRESENT MARKETING SYSTEM

Indian system of agricultural marketing suffers from a number of defects. As a consequence, the Indian farmer is deprived 'of a fair price for his produce. The main defects of the agricultural marketing system are discussed here.

1. Improper warehouses

There is an absence of proper ware housing facilities in the villages. Therefore, the farmer is compelled to store his products in pits, mud-vessels, "Kutcha" storehouses, etc. These unscientific methods of storing lead to considerable wastage. Approximately 1.5% of the produce gets rotten and becomes unfit for human consumption. Due to this reason supply in the village market increases substantially and the farmers are not able to get a fair price for their produce. The setting up of Central Warehousing Corporation and State Warehousing Corporation has improved the situation to some extent

2. Lack of grading and standardization

Different varieties of agricultural produce are not graded properly. The practice usually prevalent is the one known as "dara" sales wherein heap of all qualities of produce are sold in one common lot Thus the farmer producing better qualities is not assured of a better price. Hence there is no incentive to use better seeds and produce better varieties.

3. Inadequate transport facilities

Transport facilities are highly inadequate in India. Only a small number of villages are joined by railways and pucca roads to mandies. Produce has to be carried on slow moving transport vehicles like bullock carts. Obviously such means of transport cannot be used to carry produce to far-off places and the farmer has to dump his produce in nearby markets even if the price obtained in these markets is considerably low. This is even more true with perishable commodities.

4. Presence of a large number of middlemen

The chain of middlemen in the agricultural marketing is so large that the share of farmers is reduced substantially. For instance, a study of D.D. Sidhan revealed, that farmers obtain only about 53% of the price of rice, 31% being the share of middle men (the remaining 16% being the marketing cost). In the case of vegetables and fruits the share was even less, 39% in the former case and 34% in the latter. The share of middle- men in the case of vegetables was 29.5% and in the case of fruits was 46.5%. Some of the intermediaries in the agricultural marketing system are -village traders, Kutcha arhatiyas, pucca arhatiyas, brokers, wholesalers, retailers, money lenders, etc.

5. Malpractices in unregulated markets

Even now the number of unregulated markets in the country is substantially large. Arhatiyas and brokers, taking advantage of the ignorance, and illiteracy of the farmers, use unfair means to cheat them. The farmers are required to pay arhat (pledging charge) to the arhatiyas, "tulaii" (weight charge) for weighing the produce, "palledari" to unload the bullock-carts and for doing other miscellaneous types of allied works, "garda" for impurities in the produce, and a number of other undefined and unspecified charges. Another malpractice in the mandies relates to the use of wrong weights and measures in the regulated markets. Wrong weights continue to be used in some unregulated markets with the object of cheating the farmers.

6. Inadequate market information

It is often not possible for the farmers to obtain information on exact market prices in different markets. So, they accept, whatever price the traders offer to them. With a view to tackle this problem the government is using the radio and television media to broadcast market prices regularly. The news papers also keep the farmers posted with the latest changes in prices. however the price quotations are sometimes not reliable and sometimes have a great time-lag. The trader generally offers less than the price quoted by the government news media.

7. Inadequate credit facilities

Indian farmer, being poor, tries to sell off the produce immediately after the crop is harvested though prices at that time are very low. The safeguard of the farmer from such "forced sales" is to provide him credit so that he can wait for better times and better prices. Since such credit facilities are not available, the farmers are forced to take loans from money lenders, while agreeing to pledge their produce to them at less than market prices. The co-operative marketing societies have generally catered to the needs of the large farmers and the small farmers are left at the mercy of the money lenders.

Thus it is not possible to view the present agricultural marketing system in India in isolation of (and separated from) the land relations. The regulation of markets broadcasting of prices by All India Radio, improvements in transport system, etc., have undoubtedly benefited the capitalist farmers, and they are now in a better position to obtain favourable prices for their "market produce" but the above mentioned changes have not benefited the small and marginal farmers to any great extent.

CHARACTERISTICS OF AGRICULTURAL PRODUCT

Agricultural products differ in nature and contents from industrial goods in the following respects.

1. Agricultural products tend to be bulky and their weight and volume are great for their value in comparison with many industrial goods.
2. The demand on storage and transport facilities is more heavy, and more specialized in case of agricultural products than in the case of manufactured commodities.
3. Agricultural commodities are comparatively more perishable than industrial goods. Although some crops such as rice and paddy retain their quality for long time, most of the farm products are perishable and cannot remain long on the way to the final consumer without suffering loss and deterioration in quality.
4. There are certain agricultural products such as mangoes and grapes which are available only in their seasons but this condition of seasonal availability is not found in the case of industrial goods.
5. Agricultural produce is to be found scattered over a vast geographical area and as such its collection poses a serious problem. But such is not condition in the case of industrial goods.
6. There are various kinds and varieties in farm produce and so it is difficult to grade them.
7. The farmers especially in countries like India has low holding-back. Therefore he has to sell his produce immediately after the harvest at whatever price he can fetch because of his pressing needs.
8. Finally, both demand and supply of agricultural products i are inelastic. A bumper crop, without any minimum guaranteed support price from the government may spell disaster for the farmer. Similarly the farmer may not really be in a position to take advantage of shortages or deficit crop. These benefits may pass on only to the middleman.

Methods of Sale and Marketing Agencies

The marketing of agricultural produce is generally transacted in one of the following ways.

1. Under cover or the Hatta System: Under this system, the sale is effected by twisting or clasping the fingers of the sellers agent under cover of a cloth. The cultivator is not taken into confidence until the final bid is cleared.
2. Open auction system: Under this system the agent invites bids for the produce and to the highest bidder the produce is sold.
3. Dara system: Another related system is to keep the heaps of grains of different quantities and sell them at fiat rates without indulging in weightment etc.
4. Moghum sale: Under this system, sale is based on the verbal understanding between buyers and sellers and without mentioning the rate as it is understood that the buyers will pay the prevailing rate.
5. Private agreement: He seller may invite offers for his produce and may sell to one who might have offered the highest price for the produce.
6. Government purchase: The government agencies lay down fixed prices for different qualities of agriculture commodities. the sale is effected after a gradual processing for gradation and proper weightment. This practice is also followed in co-operative and regulated markets.
7. Marketing agencies: The various agencies engaged in the marketing of agricultural produce can be classified into two categories, viz., (i) government and quasi private agencies like the co-operative societies and (ii) private agencies. A chain of middlemen may be found operating both in Government and private agencies. The more important among these are as follows:-

a. Merchant is the most usual purchaser of the produce, he deals in his individual capacity.

b. Itinerant Beoparis (merchants) visit different villages, collect the produce, and take to the nearest market.

c. Take the weighing men from the villages to the dealers in town.

d. Agents are concerned with the assembling and distribution of agricultural produce.

EXISTING SYSTEMS OF AGRICULTURAL MARKETING IN INDIA

The existing systems of agricultural marketing in India are as briefly described here.

1. Sale to moneylenders and traders

A considerable part of the total produce is sold by the farmers to the village traders and moneylenders. According to an estimate 85% of wheat, 75% of oil seeds in U.P., 90% of jute in West Bengal and 60% of wheat, 70% of oil seeds and 35% of cotton in Punjab are sold by the farmers in the villages themselves. Often the money lenders act as a commission agent of the wholesale trader.

2. Hats and shanties

Hats are village markets often held once or twice a week, while shanties are also village markets held at longer intervals or on special occasions. The agents of the wholesale merchants, operating in different mandies also visit these markets.

The area covered by a "hat" usually varies from 5 to 10 miles. Most of "hats" are very poorly equipped, are uncovered and lack storage, drainage, and other facilities. It is important to observe that only small and marginal farmers sell their produce in such markets. The big farmers with large surplus go to the larger wholesale markets.

3. Mandies or wholesale markets

One wholesale market often serves a number of villages and is generally located in a city. In such mandies, business is carried on by arhatiyas. The farmers sell their produce to these arhatiyas with the help of brokers, who are generally the agents of arhatiyas. Because of the malpractices of these middlemen, problems of transporting the produce from villages to mandies, the small and marginal farmers are hesitant of coming to these mandies.

The arhatiyas of these mandies sell off the produce to the retail merchants. However, paddy, cotton and oilseeds are sold off to the mills for processing. The marketing system for sugarcane is different. The farmers sell their produce directly to the sugar mills.

4. Co-operative marketing

To improve the efficiency of the agricultural marketing and to save farmers from the exploitation and malpractices of middlemen, emphasis has been laid on the development of co-operative marketing societies. Such societies are formed by farmers to take advantage of collective bargaining. A marketing society collects surplus from it members and sell it in the mandi collectively. This improves the bargaining power of the members and they are able to obtain a better price for the produce. In addition to the sale of produce, these societies also serve the members in a number of other ways.

IMPROVEMENT OF AGRICULTURAL MARKETING SYSTEM

Government of India has adopted a number of measures to improve agricultural marketing, the important ones being - establishment of regulated markets, construction of warehouses, provision for grading, and standarization of produce, standarisation of weight and measures, daily broadcasting of market prices of agricultural crops on All India Radio, improvement of transport facilities, etc.

1. Marketing surveys

In the first place the government has undertaken marketing surveys of various goods and has published these surveys. These surveys have brought out the various problems connected with the marketing of goods and have made suggestions for their removal.

2. Grading and standardization

The government has done much to grade and standardize many agricultural goods. Under the Agricultural Produce (Grading and Marketing) Act the Government has set up grading stations for commodities like ghee, flour, eggs, etc. The graded goods are stamped with the seal of the Agricultural Marketing Department -AGMARK The "Agmark" goods have a wider market and command better prices. A Central Quality Control Laboratory has been set up at Nagpur and eight other regional laboratories in different parts of the country with the purpose of testing the quality and quality of agricultural products applying for the Government's "Agmark" have been created The Government is further streamlining quality control enforcement and inspection and improvement in grading.

3. Organization of regulated markets

Regulated markets have been organized with a view to protect the farmers from the malpractices of sellers and brokers. The management of such markets is done by a market committee which has nominees of the State Government, local bodies, arhatiyas, brokers and farmers. Thus all interests are represented on the committee. These committees are appointed by the Government for a specified period of time. Important functions performed by the committees can be summarized as follows.

a. fixation of charges for weighing, brokerages etc.,
b. prevention of unauthorized deductions, underhand dealings, and wrong practices by the arhatiyas,
c. enforcing the use of standardized weights,
d. providing up to date and reliable market information to the farmers, and
e. settling of disputes among the parties arising out of market operations.

The system of regulated markets has been found to be very useful in removing fraudulent practices followed by brokers and commission agents and in standardizing market practices. The committee is responsible for the licensing of brokers and weightmen. It is nested with powers to punish anyone who is found guilty of dishonest and fraudulent practices. 1t is the policy of the government to convert all markets in the country into the regulated type.

Regulated markets aim at the development of the marketing structure to have the following.

1. ensure remunerative price to the producer of agricultural commodities,
2. reduce non functional margins of the traders and commission agents, and
3. narrow down the price spread between the producer and the consumer.

To achieve these objectives, the government would go in for comprehensive and rapid expansion of regulated marketing systems. The success achieved in states like Punjab and Haryana, where regulated markets have been established in major producing areas with linked up satellite markets in the rural growth centres would be aimed at, in other areas where intensive production is taken up. The regulating marketing system has also proved a good source of generating income for the marketing boards and for use in rural infrastructure. The regulated market complex will also include facilities for grading and for monitoring of prices.

The development of regulated markets is proposed especially in areas where commercial crops like cotton, jute, tobacco and important non-traditional crops are produced and sold in weekly markets and hats. Co-operative marketing and distribution and banking will also be linked with the regulated markets. These markets will cover all the major crops. Separate market yards are proposed for livestock, fish, fruits and vegetables.

There are now over than 6,050 regulated markets with the establishment of these regulated markets. The malpractices in mandies have disappeared and the market charges have been rationalized. As much as 70% of agricultural produce is now sold in regulated markets.

In this connection, the steps taken to standardize the weight and measures in the country should be mentioned. The government has successfully replaced the different systems of weights and measures prevalent in the country with the metric system.

4. Provision of warehousing facilities

To prevent distress sale by the farmers, particularly the small and marginal farmers, due to prevailing low prices, rural go downs have been set up. The government has done much to provide warehousing in towns and villages.

The Central Warehousing Corporation was set up in 1957 with the purpose of constructing and running go downs and warehouses for the storage of agricultural produce. The states has set-up the State Warehousing Corporations with the same purpose. At present the Food Corporation is constructing its own network of go downs in different parts of the country. The total storage capacity in the country was 27 million tonnes at the end of the sixth plan.

5. Dissemination of market information

The government has peen giving attention to the broadcasting of market information to the farmers. Since most villages have radio sets, these broadcasts are actually heard by farmers. The newspapers also publish agricultural prices either daily or weekly accompanied by a short review of trends.

6. Directorate of marketing and inspection

The directorate was set up by the Government of India to co-ordinate the agricultural marketing of various agencies and to advise the Central and State Governments on the problems of agricultural marketing. Activities of this directorate includes the following.

a. promotion of grading and standardization of agricultural and allied commodities;

b. statutory regulation of markets and market practices;

c. training of personnel;

d. market extension;

e. market research, survey and planning and

f. administration of O>ld Storage Order, 1980 and Meat Food Products Order, 1973.

The directorate has so far formulated grade specification for 142 agricultural commodities. It enforces compulsory quality control before export on as many as 41 agricultural commodities. It is extending financial assistance to selected regulated markets for providing grading facilities for important commodities like tobacco, jute, cotton, groundnut and cashew nut at the producers level.

An allied task is the one related to marketing research and survey. This should aim at determination of best handling methods of produce to minimize losses, damage and costs, improved methods of wholesaling and retailing and planning for new marketing facilities at appropriate centres. With this aim in view, the Directorate is currently implementing two schemes.

- Market research and planning.
- Market planning and design.

Under the former scheme, the Directorate has been carrying out country-wide marketing surveys on live-stock and important agricultural and horticultural commodities to identify and study the problems of agricultural marketing. Under the latter scheme, the Directctorate has set up a Marketing Planning and Design Centre at Faridabad and a training centre and Workshop at Nagpur to study the packaging grading and marketing of selected fruits and vegetable and also advise the authorities on the designing of fruits and vegetable markets.

7. Government purchases and fixation of support prices

In addition to the measures mentioned above, the Government also announces minimum support price for various agricultural commodities from time to time in a bid to ensure fair returns to the farmers. These prices are fixed in accordance with the recommendations of the Agricultural ,Price Commission. If the prices start falling below the declared level (say, as a result of glut in the market), the Government agencies like the Food Corporation of India intervene in the market to make direct purchase from the farmers at the support prices. These purchases are sold off by the Government at reasonable price through the public distribution system.

COOPERATIVE MARKETING

Though the above measures have improved the system of agricultural marketing to some extent, a major part of the benefits has been derived by large farmers, who have adequate marketable surplus. However, the small and marginal farmers continue to sell a major part of their produce to moneylenders to meet their credit needs and these moneylenders offer them very low prices. Therefore it is essential to form cooperatives of the small and marginal farmers to enable them to obtain fair prices for their produce. The advantages that co-operative marketing can confer on the farmer are multifarious, some of which are listed below.

1. Increases bargaining strength of the fanners

Many of the defects of the present agricultural marketing system arise because often one ignorant and illiterate farmer (as an individual) has to face well-organised mass of clever intermediaries. If the farmers join hands and for a co-operative, naturally they will be less prone to exploitation and malpractices. Instead of marketing their produce separately, they will market it together through one agency.

2. Direct dealing with final buyers

In cases, the co-operatives can altogether skip the intermediaries and enter into direct relations with the final buyers. This practice will eliminate exploiters and ensure fair prices to both the producers and the consumers.

3. Provision of credit

The marketing co-operative societies provide credit to the farmers to save them from the necessity of selling their produce immediately after harvesting. This ensures better returns to the farmers.

4. Easier and cheaper transport

Bulk transport of agricultural produce by the societies is often easier and cheaper. Sometimes the societies have their own means of transport. This further reduces cost and botheration of transporting produce to the market.

5. Storage facilities

The co-operative marketing societies generally have storage facilities. Thus the farmers can wait for better prices. Also there is no danger to their crop yield from rains, rodents and thefts.

6. Grading and standardization

This task can be done more easily for a co-operative agency than for an individual farmer. For this purpose, they can seek assistance from the government or can even evolve their own grading arrangements.

7. Market intelligence

The co-operatives can arrange to obtain data on market prices, demand and supply and other related information from the markets on a regular basis and can plan their activities accordingly.

8. Influencing marketing prices

While previously the market prices were determined by the intermediaries and merchants and the helpless farmers were mere spectators force to accept, whatever was offered to them, the co-operative societies have changed the entire complexion of the game. Wherever strong marketing co-operative are operative, they have bargained for and have achieved, better prices for their agricultural produce.

9. Provision of inputs and consumer goods

The co-operative marketing societies can easily arrange for bulk purchase of agricultural inputs, like seeds, manures fertilizers etc. and consumer goods at relatively lower price and can then distribute them to the members.

10. Processing of agricultural produce

The co-operative societies can undertake processing activities like crushing seeds, ginning 'and pressing of cotton, etc.

In addition to all these advantages, the co-operative marketing system can arouse the spirit of self-confidence and collective action in the farmers without which the programme of agricultural development, howsoever well conceived and implemented, holds no promise to success.

WAREHOUSING IN INDIA

Warehousing facilities are necessary to prevent the loss arising out of defective storage and also to equip the farmers with a convenient instrument of credit. Both the Agricultural Finance Sub-Committee (1945) and the Rural Banking Enquiry Committee (1950) emphasized the importance of ware housing as a method of promoting rural banking and finance in India.

All India Rural Credit Survey Committee (1954) recommended a three tire system of warehousing: at the national level, state and district level, village and rural level.

At present there are three main agencies in the public sector which are engaged in building large scale storage/warehousing capacity. They are: the Food Corporation of India (FCI), Central Warehousing Corporation (CWC), and State Warehousing Corporation (SWC).

FCI provides storage capacity for food grains. It has its own go downs and it also hires storage capacity from other sources such as CWC, SWC's, State Governments and private parties. In 1960-61, there were only 40 general warehouses in the countries with a total capacity of less than 0.1 million tonnes. By the end of 1988-89, the three public sector units has a storage capacity of nearly 32 million tonnes.

Besides, public sector agencies, co-operatives have also constructed warehouses in rural areas for storage of their members' produce, for stocking of fertilizers and other inputs and consumer articles. To avoid unfair competition with the go downs of the co-operative marketing societies, the state warehousing corporations do not open warehouses at any place below the sub divisional level. By 1987-88, a total storage capacity of over 10 million tonnes in the co-operative sector was available.

Kinds of warehouses

There are broadly speaking four kinds of warehouses.

They are:

a. private warehouses which are usually maintained by joint-stock companies, firms and individuals;

b. Duty-paid public ware houses which are maintained by dock authorities or port trust authorities at port

c. Bonded warehouses which are maintained either by dock authorities or by the Government and

d. licensed warehouses which are private warehouses run by co-operative societies or by private agencies, after obtaining license from the Government.

Benefits of the warehouses

The following are some of the benefits of the warehouses.

a. It gives withholding power to the agriculturist to tide over difficulties and helps them to secure better prices for their produce.

b. It gives purchasing power to traders.

c. It tends to cushion the price fluctuation and stabilize prices as it equates supply to demand.

d. It facilitates future trading.

e. It plays a very important role in implementing the agricultural price policy of the Government

f. It obviates the need for unnecessary cross-transport.

g. Huge wastages which occur owing to improper storage of agricultural produce will be minimized if warehousing develops on a large scale.

h. Warehouses render various subsidiary services, such as sorting and packing commodities for shipment, cleaning and drying goods and preparing them for the market, acting as forwarding agents for exporters of good, purchasing goods on behalf of clients, and collecting and disseminating marketing intelligence.

Progress

Storage and warehousing facilities for agricultural crops on a commercial basis are available both in the public and the private sectors. The public sector dominates and accounts for a significantly large share of the total capacity available in the economy. The main institutional agencies providing these facilities are summarized here as Central and State Warehousing Corporations, the Food Corporation of India and the cooperatives.

The Central Warehousing Corporation was set up in 1962 to do the following functions:

a. acquire and build go downs and warehouses at suitable places,

b. arrange facilities for the transport of agricultural produce .and inputs,

c. subscribe to the share capital of State Warehousing Corporations, and

d. act as agent of the Government for the purpose of purchase, sale, storage and distribution of agricultural produce and crops.

The State Warehousing Corporations have been set-up at places of state and district importance. Their functions are the same within a state as those of CWC at the national level.

IDEAL MARKETING SYSTEM

The ideal marketing system is one that maximizes the long run welfare of society. To do this, it must be physically efficient, otherwise the same output could be produced with fewer resources, and it must be electively efficient, otherwise a change in allocation could increase the total welfare and where income distribution is not a consideration.

For maximum physical efficiency, such basic physical functions as transportation, storage, and processing should be carried on in such a way so as to achieve the highest output per unit of cost incurred on them. Similarly an ideal marketing system must allocate agricultural products in time, space and form to intermediaries and consumers in such proportions and at such prices as to ensure that no other allocation would make consumers better off. To achieve this condition, prices throughout the marketing system must be efficient and must at the same time be equal to the marginal costs of production and marginal consumer utility.

The following characteristics should exist in a good marketing system.

1. There should not be any government interference in free and market transactions. The method of intervention include, restrictions on food

grain movements, restrictions on the quantity to be processed, or on the construction of processing plant, price supports, rationing, price ceiling, entry of persons in the trade, etc. When these conditions are violated, the inefficiency in the market system creeps in and commodities pass into the black market. They are not then easily available at the fair prices.

2. The marketing system should operate on the basis of the independent, but systematic and orderly, decisions of the millions of the individual consumer and producers whose lives are affected by it.
3. The marketing system should be capable of developing into an intricate and far-flung marketing systems in view of the rapid development of the urban industrial economy.
4. The marketing system should bring demand and supply together and should establish an equilibrium between the two.
5. The marketing system should be able to generate employment by ensuring the development of processing industries and convincing the people to consume more processed foods, consistent with their tastes, habits and income levels.

Scientific Marketing of Farm Products

The tendency among the farmers to market their produce has been increasing. Production is complete only when the produce is marketed at a price remunerative to the farmer. Increasing specialization in production of higher marketable/ marketed surplus of the produce and alternative channels of marketing have increased the importance of the marketing activity for the farmers. However, marketing activity should be guided by certain basic principles which are briefly explained. The farmers can gain more if they follow the following principles of scientific marketing.

1. Always bring the produce for sale after cleaning it: Impurities, when present, lower the price offered by the traders-buyers in the market. The fall in price is more than the extent of impurity present in the produce would warrant. Clean produce attracts more buyers.
2. Sell different qualities of products separately: The produce of different varieties should be marketed separately. It has been observed that when different varieties of products are marketed separately, the farmers get a higher price because of the buyers preference for specific varieties.
3. Sell the produce after grading it: Graded produce is sold off quickly. The additional income generated by the adoption of grading and standardization

is more than the cost incurred in the process of grading and standardization. This shows that there is an incentive for the farmers for the production of good quality products.

4. Keep abreast of market information: Price information helps him to take decisions about when and where to sell the produce, so that a better price may be obtained.
5. Carry bags/packs of standard weights: Farmers should weigh their produce and fill each bag with a fixed quantity. Majority of the farmers do not weigh their produce before taking it for sale and suffer loss by way of a possible malpractice in weighing, or they may have to make excess payments in transit (octroi, transport costs, etc.).
6. Avoid immediate post-harvest sales: The prices of the produce touch the lowest level in the peak marketing season. Farmers can get better prices by availing of warehouses facilities existing in their areas. Farmers can meet their cash needs by pledging the warehouse receipt to nationalized banks.
7. Patronize co-operative marketing societies: Farmers can get better prices by sales through a cooperative and marketing society and can avoid the possibility of being cheated. The cost of marketing particularly the transportation cost for farmers having a small quantity of marketable surplus, is minimized, for transportation is arranged co-operatively by the society and the profit earned by the society is shared among its members.
8. Sell the produce in regulated markets: The farmers should take their produce for sale to the nearly regulated markets rather than sell them in village or unregulated markets. In regulated markets marketing charges are on very few items. They get the sales slips in the regulated markets, which show the quantity of the produce marketed and the amount of charges deducted from the values of the produce. Sales slips protect farmers against the malpractices of deliberate erroneous accounting or unauthorized deductions.

COMMODITY MARKETING

The term 'commodity' is commonly used in reference to basic agricultural products that are either in their original form or have undergone only primary processing. Examples include cereals, coffee beans, sugar, palm oil, eggs, milk, fruits, vegetables, beef, cotton and rubber. A related characteristic is that the production methods, postharvest treatments and/or primary processing to which they have been subjected, have not imparted any distinguishing characteristics or attributes. Thus, within a particular grade, and with respect to a given variety, commodities coming from different suppliers, and even different countries or continents, are ready substitutes for one another. For example whilst two varieties of coffee bean, such as robusta and arabica, do have differing characteristics but two robustas, albeit from different continents, will, within the same grade band, have identical characteristics in all important respects. Agricultural commodities are generic, undifferentiated products that, since they have no other distinguishing and marketable characteristics, compete with one another on the basis of price.

Agricultural products whose production methods, postharvest treatments and/or primary processing have not imparted any distinguishing characteristics or attributes. Within a particular grade, and with respect to a given variety, commodities coming from different suppliers, and even different countries or continents, are ready substitutes for one another. This chapter describes commodity marketing processes. Five categories of commodity are discussed: grains, livestock and meat, poultry and eggs and fresh milk. A commodity marketing system encompasses all the participants in the production, processing and marketing of an undifferentiated or unbranded farm product, including farm input suppliers, farmers, storage operators, processors, wholesalers and retailers involved in the flow of the commodity from initial inputs to the final consumer. The commodity marketing system also includes all the institutions and arrangements that effect and coordinate the successive stages of a commodity flow such as the government and its parastatals, trade associations, cooperatives, financial partners, transport groups and educational organizations related to the commodity. The commodity system framework includes the major linkages that hold the system together such as transportation, contractual coordination, vertical integration, joint ventures, tripartite marketing arrangements, and financial arrangements. The systems approach emphasises the interdependence and inter relatedness of all aspects of agribusiness, namely : from farm input supply to the growing, assembling, storage, processing, distribution and ultimate consumption of the product. The stages of commodity marketing are: assembly, transportation, storage, grading and classification, processing, packaging and distribution and retailing.

The principal participants in grain marketing are producers, marketing boards, grain elevators, brokers, millers, livestock farmers, animal feed processors, millers, other food manufacturers, grain exchanges and exporters. A common feature of grain marketing systems is the co-existence of government marketing agencies and a parallel private marketing channel. In the post-market liberalisation era many of these parastatals have either been disbanded or have been assigned a specialised role such as acting as the buyer-of-last-resort or maintaining food security reserves.

Whether storage takes place on the farm or in silos off the farm, increases in the value of products due to their time utility must be sufficient to compensate for costs at this stage, or else storage will not be profitable. Under market liberalisation, some grain marketing parastatals have suffered substantial falls in the volumes of grains which they are handling and a number of marketing parastatals now rent some of their storage capacity to farmers, grain traders and other participants in the grain marketing system. Two types of storage facility, are commonly found: the bulk storage facility where cereals are stored in concrete and/or metal bins, and the bag storage facility where the crop is stored under tarpaulin sheets. Grading usually takes place in the grain store. Grades and standards contribute to operational and pricing efficiency by providing buyers and sellers with a system of communicating price and product information. However, grading standards for domestic markets must originate from the industry itself as and when it becomes apparent that the consumer is willing to pay a higher average price for the sorted product.

Amongst the major challenges to grain marketing systems in developing countries are: rationalisation of storage facilities in terms of their location, readjustment of farmer prices to accurately reflect production and marketing costs,

Attempts by livestock farmers to adjust production to the prevailing demand are often frustrated by the lags, between changes in demand and adjustments to supply. Meat prices increase as producers withdraw animals in the short term to build up supplies in the long term. Conversely, when prices fall and farmers selling off animals to reduce production levels the act of doing so further depresses prices.

Grains for human consumption and those for animal feeds have a direct bearing upon one another. Moreover, the price of grains converted to feed will influence the costs of meat production and, therefore, the price of meat. These interrelationships are as important to government policy makers as they are to the participants in the livestock and meat marketing systems.

The major players in livestock marketing are rural traders, local cooperatives, order buyers, commissioning agents, auctions, terminal public markets and meat packers. The systems of grading beef cattle and carcasses vary by culture but may include carcass shape, fat content and meat yield. The assessment of these and other criteria is usually subjective.

The income elasticity of demand for beef in particular, and other meats in general, tends to be strongly positive in LDCs. As per capita incomes rise in developing countries, the demand for the principal livestock products meat, milk and eggs also rises. However the development of livestock industries in developing countries is sometimes constrained by the form of traditional techniques of food preparation and shopping behaviour. In many cultures it is customary to cook meat for a very long time, especially if the favoured dishes are curries and stews and in these circumstances little interest is shown in paying premium prices for more expensive cuts of meat and butchers have no incentive to add value to the product by dividing the carcass into various consumer cuts.

The poultry and egg sector is highly fragmented in most developing countries. The principal external features of an egg, which collectively determine its quality are shell texture, colour, shape and condition. The internal condition of an egg can be established using the non-destructive tests of candling. Of particular importance are the position of the yolk within the shell, yolk colour; the extent of blood spots and the translucence and firmness of the albumen and the size of the air cell inside the shell.

The weight of a poultry carcass is a primary attribute when grading the bird. The weight will vary by breed, sex, age and in accordance with the feeding regime of the bird. The eating quality of poultry meat is expressed in terms of meat tenderness, juiciness flavour and skin colour. Poultry is less of a commodity than it was at one time. There is a close relationship between beef and milk production. In many countries, beef production is subsidiary to dairy production with sometimes over two-thirds of cattle sold for beef being animals culled from the dairy herd. The breeds and production methods which give the best milk yields rarely give corresponding results in terms of beef production; or vice versa. Milk production is seasonal with peaks just after the rains when there is lush pasture available and a progressive declines the further into the dry season. Assembly is simple and direct with milk going direct from diary farms to the processing plant. Milk collection is usually undertaken by either a marketing board or a cooperative. In general fluid milk attracts higher prices than milk destined for use in processed products. This is due to the additional costs of marketing a highly perishable product and moving and storing a very bulky commodity. The second explanatory factor is the fact that fluid milk has a lower elasticity of demand than do processed products.

Stages in a commodity marketing system

A commodity marketing system encompasses all the participants in the production, processing and marketing of an undifferentiated or unbranded farm product (such as cereals), including farm input suppliers, farmers, storage operators, processors, wholesalers and retailers involved in the flow of the commodity from initial inputs to the final consumer. The commodity marketing system also includes all the institutions and arrangements that effect and coordinate the successive stages of a commodity flow such as the government and its parastatals, trade associations, cooperatives, financial partners, transport groups and educational organisations related to the commodity. The commodity system framework includes the major linkages that hold the system together such as transportation, contractual coordination, vertical integration, joint ventures, tripartite marketing arrangements, and financial arrangements. The systems approach emphasises the interdependence and inter relatedness of all aspects of agribusiness, namely : from farm input supply to the growing, assembling, storage, processing, distribution and ultimate consumption of the product.

Grain marketing

The principal participants in grain marketing systems are producers, marketing boards, grain elevators, brokers, millers, livestock farmers, animal feed processors, millers, other food manufacturers, grain exchanges and exporters.

The physical marketing system begins with the assembling and collecting points located in the rural areas close to the producers. The next stage involves the storage areas at the national grains marketing facilities owned and operated by an appointed parastatal and/or private grain elevator; and the grain milling companies which in some countries are privately owned and in others are government enterprises. Although the size and methods of operation differ from country to country, the local assembling and collection points usually have grains brought to them either directly by the farmer-producers themselves or by rural entrepreneurs. Thus in the case of grain, the assembly and storage functions are typically combined at this marketing stage. In countries where a marketing parastatal has been given a monopoly in grain trading, private traders are sometimes authorised to buy grains from farmers (i.e. local buying agents) on behalf of the parastatal.

A common feature of grain marketing systems is the co-existence of a government marketing agency (parastatal) and a parallel private marketing channel with myriad's of private traders. Public grain marketing agencies are government appointed parastatals assigned to control or regulate the system. Prior to market liberalisation marketing parastatals were recognised, in many developing countries, as the "official channel" of the maize marketing system, and were responsible for setting the prices for major cereals (i.e. producer, into-depot, ex-depot, into-mill, ex-mill and consumer or retail). Depending on the country, these agencies usually consisted of one or more ministries related to cereals production and influencing public policies affecting food production and consumption and a parastatal established to operationalise the regulatory provisions of the public policies enacted for a given crop. In many developing countries parastatals remain important players in the grain marketing system even though their role may have changed. In the post-market liberalisation era many, but not all, of these parastatals have either been disbanded or have been assigned a specialised role such as acting as the buyer-of-last-resort or maintaining food security reserves.

The second class of actor in the commodity marketing system is private agents. These include private individuals operating in the system as petty assemblers, traders large-scale merchants, millers (both large corporations and small rural operators), brokers and retailers of grain products. The exact quantity of grain flowing through the private channel is often not known. However, most of the agents in the channel operate in rural areas and penetrate the remotest areas of the rural areas to purchase grains. A common justification for establishing parastatals is, that parastatals get to hinterlands that private operators cannot or will not reach. However, lkpi2 claims that it is private marketing agents that more often get to remote hinterlands to buy and collect maize from farmers when the government agencies fail. However, in those developing countries where national structural adjustment programme have not yet been initiated, inter-provincial grain movement controls are so strictly policed that private agents in the system cannot legally and profitably transport grain from one production zone to another. For those countries that are already adjusting their economies structurally, market liberalisation is making or will soon make much restrictions on commodity movement irrational and new form of intermediary are coming into being. Among these are commissioning agents or brokers. These entrepreneurs do not take title to the grain but take responsibility for selling the grain. They act as agents for the grain seller who may be a farmer, a grain trader or grain elevator.

Grain storage

Whether storage takes place on the farm or in silos off the farm, increases in the value of products due to their time utility must be sufficient to compensate for costs at this stage, or else storage will not be profitable. These costs will include heating, lighting, chemical treatments, store management and labour, capital investment in storage and handling equipment, interest charges and opportunity costs relating to the capital tied up in stocks. Among the less tangible costs is the risks attached to storage. These include shrinkage due to pilferage, pests, fungal growths and loss of quality due to ageing. Another risk is that demand could fall with adverse effects on prices.

Since the advent of structural adjustment programmes and market liberalisation, some grain marketing parastatals have lost their monopoly of the market and consequently the volumes of grains which they are handling has dropped substantially. This means that they no longer require all of their storage capacity and a number of marketing parastatals now rent some of their storage capacity to farmers, grain traders and other participants in the grain marketing system.

Two types of storage facility, are commonly found, namely: the bulk storage facility where cereals are stored in concrete and/or metal bins, and the bag storage facility where the crop is stored either inside a warehouse or in the open and then covered by tarpaulin sheets. In comparative terms, the advantage of a bulk over a bag storage system are that it is more efficient because it:

- Reduces congestion at the depots by not allowing for bagged maize to be dumped all over the depot yard
- Reduces handling costs
- Saves foreign exchange on bags, tarpaulin and fumigation, and
- Lowers storage losses.

Its disadvantages are that:

- The initial invest is high, with a significant foreign currency component
- It is inflexible in terms of not being easily expandable to cope with changes in intake and off-take levels
- It relies heavily on an efficient transport system because a silo complex is only economically viable when throughout is at least 1.6 times its capacity3, and
- It needs skilled manpower to run and maintain the entire system.

On the othe®r hand, a bag system ideally overcomes the problems associated with a bulk system as enumerated above. Its main disadvantages are:

- Higher quality losses in storage due to insect pests and rodents
- Higher demand for foreign exchange associated with the purchase of bags
- Tarpaulin and fumigation sheets.

The depot manager controls the day-to-day operations of the depot. The duties of the depot manager include accepting, checking the quality, recording the quantity and storing the produce brought in by the farmers. Each depot's record of accepted and stored produce are then sent to headquarters for necessary action. With this information to hand, inter-depot or inter-district transfers can be effected.

Grading of grains

It is important to have a grading system which accurately describes products in a uniform and meaningful manner. Grades and standards contribute to operational and pricing efficiency by providing buyers and sellers with a system of communicating price and product information. By definition, commodities are indistinguishable from one another. However, there are differences between grades and this has to be communicated to the market. By the same measure, buyers require a mechanism to signal which grades they are willing to purchase and at what premium or discount. Prices vary among the grades depending upon the relative supply of and demand for each grade. Since the value of a commodity is directly, affected by its grade, disputes can and do arise. In fact, the government may establish grading services to serve as a disinterested third party.

Grading typically occurs at the assembly stage or when a product moves into storage, during storage, or just before it leaves storage. Grading is not normally a separate marketing stage, although it has been separately identified in table 6.1 in this chapter. It is a function provided by the storage firm or the commodity merchant or the government. Prescribed procedures for grading are set forth by the trade members of commodity markets or else are stated in governments regulations. Grading may be undertaken by a member of the trade specialising in a particular commodity. Several lots of grain, oilseeds, and cotton are combined to produce a grade level required for a particular sale. This gives rise to what are known as house grades. A merchant's primary marketing advantage may be a reputation for house grades of consistent quality.

The absence of grades and standards restricts the development of effective and efficient marketing systems. For example, for some time the government of Nepal have been trying to establish an internal food marketing system. It has no nationally integrated market but due to its topography and poor communication routes there is little inter-state trade. Instead each state has closer trade relations with neighbouring Indian states. There are substantial obstacles to achieving an integrated national market as explained by Chong Yeong Lee4;

"The lack of a unified measurement system also hampers the development of marketing. In Tarai, grain is measured by weight (1 mand=37.3kg) and in the Hills and Inner Valleys it is measured by volume (1 pathi = 4.54 litres). The same unit of weight, such as "seer", is equivalent to 0.93kg of meat in the Tarai area, 0.79kg in Kathmandu and 0.25kg in Pokhara."

Clearly, the confusion which this situation gives rise to, mitigates against the successful development of a national market. Moreover, effective standardisation is basic to an efficient pricing process. Consumers use the price differentials they are willing to signal to suppliers what they want with regard to produce quantities and qualities. If produce is not in well defined units of quantity and quality then the pricing mechanism fails as a device for communicating consumer wants to suppliers. However, Dixie5 warns that any grading standards for domestic markets must originate from the industry itself as and when it becomes apparent that the consumer is willing to pay a higher average price for the sorted product. He points out that,

"Although national standards can probably be justified for export, when compulsory minimum standards are introduced for the home market it will put up prices to the consumer", and that this would,"...lower consumption and reduce the size of the local market".

Usually samples of different sizes (depending on the size of the load) are taken from each lot delivered to the depot of the buyer and these are tested for compliance with the acceptance standards. The results determine the grade into which the whole lot from which the sample came is classified to determine the price to be paid to the grower. Typical variables used in grading grain include:

- The moisture content of the grain
- The percentage of broken kernels
- Degree of discolouration in the grain
- The percentage of material other than grain (MOG) in a sack or load

In Sub-Saharan Africa, for example, maize is received in 4 grades, A to D, with A being the highest grade and D the lowest. The system allows for the farmer to dispute the grade awarded to his crop. In such cases, the farmer has to submit a written request within a specific time to the depot through which his crop was delivered.

Typically, after grading is effected, the farmer is issued with a receipt showing details of the type, quantity and grade of the maize delivered. The receipt constitutes an acknowledgment by the agency that it is indebted to the farmer up to the value of the crop detailed on it. From that point on, the marketing risk attached to the crop passes from the producer to the agency/parastatal. It is relevant to point out that it usually costs the same amount to process a grain receipt for 1 tonne as it does for 1,000 tonnes.

Grain processing

Grain processing is about the most important activity from the final consumer's stand point within the marketing chain of the crop. Grain, for human consumption, is usually milled into flour or meal. Usually two types of maize flour are processed in each country, namely: the refined, white and sifted (powdery) flour produced by industrial roller mills and the unrefined and coloured maize meal produced by hammer-mills. The highly refined grade of meal is generally consumed by urban dwellers, while the hammer-mill grain meal is a low-grade "whole maize meal" favoured by rural consumers and the urban poor.

In many countries, the milling industry is highly concentrated, usually with one major company controlling over eighty percent of the maize mill trade. For example, in Kenya, maize milling is controlled by the Mercat Group of Companies (Unga Ltd, etc) (90 percent): while in Malawi, it is GRAMIL, a subsidiary of ARMARC which processes eighty three percent of the maize in that country. In Tanzania, the National Milling Corporation, (NMC) handled about eighty five percent of all maize milled before it fell into financial difficulties in the early 1990s and was allowed to go out of business; in Zimbabwe it is the National Foods which controls eighty percent of maize milling in the country until progressive liberalisation of the market increased the level of competition during the first five years of the 1990s. Grains can also undergo secondary processing and be converted into more sophisticated products such as baked foods, breakfast cereals, baby foods, cooking oils, starches, sugars etc. A considerable amount of grain is also converted to animal feed. Due to this versatility in end use, the marketing chain for grains tend to be long and complex.

Challenges for grain marketing systems

Depot network and distribution of production problems arise because crop production is rarely evenly distributed across a country. In most countries cereals production is concentrated in one or two regions of the country, and the remainder are cereals deficit areas. In these circumstances it usually becomes rational and necessary for the government agency handling grain distribution, where there is one, to construct intake depots in the producing region(s), and storage depots in the major consumption areas. But in most instances, post-independent governments in some of these countries adopted rural income policies that made it mandatory for the whole country to be provided with government agency depots within short distances of every producer. The aim was to overcome political, social and economic injustices of the past. It was a policy which was also pursued in the hope that smallholder farmers would be attracted into the cash economy. In the end, this huge investment in infrastructure was economically unsustainable and most developing countries are now locating storage facilities on the basis of economic efficiency.

Excess stock problems tend to develop because of relatively high producer prices which the governments of the countries concerned establish in order to compensate farmers whose cost for production are high. Unfortunately, these systematic increases in the producer prices usually bring them to levels that are incompatible with international supply and demand conditions. To reduce the effect of these high producer prices some governments in LDCs introduced artificial exchange rates that later did more harm than good to the export economy of these countries. In addition, local demand was low, being limited to only urban demand, for human consumption, and for livestock feed production in rural areas.

Fluctuating grain supply problem results from the crop's susceptibility to weather conditions which determine the level of harvest and, therefore, the export market supply. With each country's domestic demand often varying along with the weather, LDC countries have been known to rapidly move from a surplus to a deficit situation. In a country with high import parity and low export parity prices, the increased risk of stockpiling seems to lower the real cost of storing the excess stocks and so causes government to become indifferent toward streamlining supply with demand requirements. Low levels of intra-regional trade between developing countries is explained by subsidies paid to producers. This masks any competitive advantage in grain production that growers in one country may have over those in neighbouring countries. Where there is intra-regional trade in grains, between developing countries it is most often based on periodic shortages arising from drought civil strife. As such, export demand is usually short-lived and often financed by donor agencies, and the exporting country cannot make any long term plans to develop the trade.

ACT OF AGRICULTURAL MARKETING

The Marketing of Agricultural Products Act of 1996 was motivated more by the repeal of statutory control measures than it was by their replacement with something better. Its authors knew more about what they did not want rather than what was needed by the sector as a whole. This factor, as well as the wholesale political changes and a racially divided industry, meant that difficult questions were either glossed over or conveniently left unanswered. Consequently the present act has revealed itself to be so diluted as to be passive, with gaping holes where strategic guidelines should have been. It is important to note however that this result is not because of the failure of its authors, as by admission they set out to draw an enabling act, which would vest the authority on the market participants themselves. They neglected to consider that pre-requisites for an act of this nature would be established and legitimate institutions (which were not there in a changing society), and democratic representation of all interests within such institutions(which was not there in a racially divided country).

- The result was a compromise act, which recognised the need for a national marketing council but failed to define its role and therefore the authority it needed to fulfil it.
- Removed statutory regulations and services but failed to set up an alternative delivery system to meet the real needs of farmers.
- Acknowledged developmental obligations but left them to market forces and the goodwill of rivals in the established sector.
- Established a the NAMC as a state organ with no real purpose by denying it any decision making or executive powers, and handing assets to the commodity trusts.
- Vested all decision-making powers in the Minister and therefore managed to create a large amount of what is basically administrative work for her office, while also concentrating state influence on the market more than ever before.

The emerging black farmers will no doubt wonder at the cruel coincidence by which government lost all its influence on the market, just as they anticipated their own place in the sun.

The process of liberalising the market also failed to provide for a tight and timely winding up of historical business. While the assets accrued under the previous act clearly were no longer the property of the producers (as with other tax provisions), a situation was created where the very same producers were called back to adjudicate over their dispersal. This has resulted in wasteful and extended conflict and litigation, while also delaying the advent of a new dispensation and the emergent farmer's turn at the trough. This dubious redistribution of what were public assets held by statutory bodies is the basis of most problems besetting the trusts. It is also one of many legal changes whose motivations seem dubious, and whose effect has been the denial of effective control over public assets for the majority.

Deregulation as a concept

There is a common but erroneous perception that agriculture has moved from a regulated to a deregulated system. Strictly speaking this is not true as there will always be a need for regulations to ensure the fair and proper conduct of participants within the industry. In reality the change has been from a state regulated to a market or self regulated system. It is this erroneous belief that the market can operate without regulations, that leads some to conclude that no dedicated authority is required to deal with marketing matters.

As long as there is a national system of regulations (whether it be state proclaimed or market determined), there will also be a need for a dedicated institute to manage and maintain it. Additionally the following characteristics inherent in markets and marketing also need continued attention:

- Since markets are all about competition, there will always be a certain amount of disputes and even conflict among the competitors. A legitimate and capable authority is therefore required to harmonise the market to keep it productive.
- Competitors by definition do not enjoy each other's successes, and unless there is a clear commercial benefit are unlikely to co-operate unless obliged to do so by an institution acting in the national rather than their selfish interests.
- The market operates according to the law of the jungle and is unlikely to adequately provide for the interests of the emerging sector, without public sector influence.
- Without statutory regulations, members need to agree on and enforce their own systems and standards without which there would be chaos. An institution is therefore required to provide both the legitimacy and support for self-regulation.
- As an industry taking its first tentative steps into free and global markets, agriculture still relies on the forced co-operation provided by statutory instruments such as levies.
- Agricultural products are related. For instance decisions taken on the marketing of animal feed impact on the beef and the poultry sectors. Government can ease, speed and even improve on the quality of interaction between the different stakeholders to the good of all.
- There are certain public good functions which the private sector is unlikely to carry out because the do not contribute directly to the producer's bottom line. labour practices, animal welfare, environmental conservation, food safety, and unfair practices are unlikely to surface unless competing producers are in dispute.
- Globalised markets require government to adopt negotiating positions, make concessions and enter agreements. The limited analytical capacity of the department and the vested interests of producers requires that government should maintain a balanced, inclusive and legitimate forum, which it can use for consultations with the industry.

- The interests of producers should be balanced against the very real concerns of both intermediate and final consumers. For instance issues around genetically modified products, the use of chemicals and the price of processing inputs should be aired and agreed on at an appropriate forum.

The role of statutory institutions

Marketing is a process that begins when investment choices are made, and carries on through production, harvesting, handling, storage, processing retailing and consumption. Marketing lends objective to all processes because there is no task in commercial agriculture that is not aimed at enhancing the profitable marketability of products. Marketing therefore has to be the underlying consideration of all agricultural development and support programmes. There is therefore a need for an independent national statutory institution representative of all major commodities groupings within the sector to carry out those functions that the market would not be able or willing to do on its own. Such an institution would act as a clearing-house for market information and government policy, while also affording the government a balanced view and an opportunity to place social issues on the agenda of agriculture's private sector. Without such a structure, government will relinquish all official contact with and influence over the sector.

The institution needs to be independent because it will be constituted mainly of commodity representatives, and has to have some statutory powers to enforce properly reached decisions. Representatives of commodity groups would be nominated to the council, and would be joined by consumer representatives, development sector representatives, and public sector representatives appointed by the Minister. As an industry-wide council, it should have twenty or more members and be balanced in terms of private and public sector interests, producers and consumer interests, and emerging versus established sector interests.

Public and emerging sector representations are critical to the democratisation of the industry, but are hindered by the lack of an organised skills base within agriculture. Other industries face similar constrictions, but have gone around them by including African business people and experts from other disciplines. We now have high profile business people in mining and banking who were not originally in these fields. Agriculture is as much a business as any other, and the Minister should therefore cast her net wider to entrepreneurs other than farmers in her representation.

An appropriate induction programme into agriculture and agricultural issues should enhance their participatory and representational skills. Their membership of both the institution and the trusts, together with sufficient remuneration should ensure their active and effective representation of public and emerging sector interests. This policy shift is probably the best way to accelerate the participation of Africans in mainstream agriculture.

While public representatives and to some extent emerging sector participants should be remunerated to ensure quality participation, representatives of commercial commodity groupings would be entitled to compensation for expenses incurred. This is in line with the practices of all forums where delegates represent their own interests. With quality representation, a clear mandate and sufficient authority; the institution will not only assume relevance in the eyes of producers and consumers but will also have the necessary influence to give effect to government policies.

While the institution would report to the Minister because of its statutory nature, it should be empowered to implement democratically reached decisions on levies and other such instruments. Being representative it should also be charged with resolving disputes, investigations and making rulings, as well as rendering advice on local and international markets. The Minister will oversee the council's operations and hear appeals against its decisions. The present arrangement is devoid of this important distinction.

Marketing development

While the established sector was nurtured under the state protection and support offered by the control boards, presently emerging farmers have nothing to bank on. As agricultural development all over the world has always taken place and continues to take place under special circumstances, it is likely that unless a conducive environment is deliberately created for the emerging sector, then very few new farmers will break into the mainstream. At the forefront of the liberalisation of the sector even the WTO recognises this fact by providing for special and differential treatment of the developing sector.

Special marketing channels; guaranteed share of government tenders; and encouragement of the established sector to enter joint marketing deals are just a few examples of how government could help. An exciting opportunity arises from the present concerns around food security. It is imperative that government should maintain (or cause to be maintained)strategic grain reserves. If such reserves could be procured from the developing sector through specialised channels, or government started issuing part of its welfare grants as food stamps to be used at accredited developing suppliers; then a few birds could be killed with one stone.

The present act makes no special provision for the unique interests of the emerging sector, and with this omission should be seen to favour the establishment.

It is important that this provision be created in the act and in the roles of both the council and the trusts. It should be a statutory provision for the use of part of the levies and trust funds towards the development of others. After-all development of presently established farmers was to a significant extent bank-rolled by their more developed colleagues and the public at large. And while some provisions are already being made in this regard, a deliberate and tighter strategy needs to be implemented by the council to this effect. Effective public and emerging sector representation will then ensure its successful implementation.

Advantages of good Marketing System

A good marketing system helps in the following ways:

Monetising the Produce: It helps the farmers to get the due value of their produce.

Demand Signal Platform: Marketing systems balances the demand with supply. The bridge in the form of agri-logistics, function to execute this purpose.

Market growth: India's agriculture is transitioning from a state of deficit towards a status of surpluses. Agri-logistics technologies have also modernised to allow farm produce to connect with demand further afield. An aware and competitive marketing system, will adjust itself to suit this transformation of Agriculture in India. Further this can open up new market avenues. The opening up of the markets to promote competition and long term growth as a desired outcome.

Capital formation and investment in technology: Marketing leads to investment in Modern Supply Chain infrastructure and Information and Technology applications to increase the Produce or enhance the value of Produce.

Effective Marketing System and Its advantages

Marketing effectiveness in a measure of expanding the market system to generate larger revenue streams and competitiveness. Market channel efficiency depends on reduction of marketing cost, adoption of technology in grading, packaging, transportation, storage, value addition, wholesaling, retailing and exploring economies of scale through aggregation. An effective Marketing system provides farmers their due share in the Production process and ensures timely delivery of goods and services.

The effectiveness of an agricultural marketing system varies depending on the situation of the target regions, consumer, product and technologies in hand.

The various advantages of Effective Marketing System are-

Gives demand signals to supply side –An Effective Marketing System uses Market Intelligence in order to predict the demand. An effective marketing system will strive to minimise the blind push into markets and instead promote a pull mode from markets.

Increase in revenue generation - A well organised marketing system will increase the sum total of revenue generated in the agricultural value chain An effective marketing system will also aim to persuade that there is more equitable sharing of the net revenue generated, among all the stakeholders in the value chain system.

Help in Market expansion - The effective marketing system will always promote the expanding of the market range of the producer/supplier and provide a choice of purchase for the Market expansion also boosts vibrancy in domestic agricultural trade.

Unified market - a well organised domestic marketing system will integrate by developing extensive connectivity across a network of demand centres and supply regions, creating uniformity in the market arena.

Increased competitiveness - Competitiveness also leads to improved resource optimisation and brings about cost efficient

Government Policy instruments in Agricultural Marketing

India's main policy goals remained with focus to attain food self-sufficiency, to ensure remunerative prices to farmers, and to maintain stable prices for consumers. To meet these goals, India continues to rely on following policy instruments:

Regulation of markets

The Agricultural Produce Market Committee (APMC) Act requires that farm produce be sold only at regulated markets through registered intermediaries. The Essential Commodities Act (ECA), allows central and state governments to place restrictions on the storage and movement of commodities deemed essential by governments. {APMC Act and recently proposed APLM Act have been discussed in this module}

Input subsidies

Concerns on food insecurity, necessitated provision of subsidies on fertilizer, electricity, fuel and irrigation. Intensive farming and production growth were motivators for policy interventions.

Minimum support prices

Government of India (GoI) supports producers by announcing minimum prices for certain commodities considered key to its agriculture. The MSP are set on recommendations of the Commission for Agricultural Costs and Prices (CACP), based on cost of production and other factors. There are other associated schemes such as the Price Stabilisation Scheme and Market Intervention Scheme that also come into play as minimum support systems for farmers.

Food subsidies

India has enacted the National Food Security Act 2013, to protect and assure low income consumers of access to affordable food.

Various Platforms for Marketing of Agricultural Produce

Agriculture is a state subject and various states have enacted their laws based on Model APMC Act 2003 and Model Rules 2007. These make Mandis or APMCs as main platform for farmers to sell their produce. Apart from that, there are some alternative marketing platforms also that can be utilised by farmers to sell their produce. These are as follows:

Direct Marketing

In direct marketing, the farmers directly sell with the produce consumers. These operate in two basic formats:

Farmers' Markets, and

Direct sourcing from farmer's field by processors (primary consumers).

These markets have helped in mitigating the problems of fragmented supply chain. Along with it, the quick movement of produce from farmer to consumer saves losses considerably. In these markets no market fee is charged but service charges are collected from sellers. About 488 such farmers' markets are operating across different States in the name of Apnamandis in Punjab, Haryana, Rythu bazaars in Andhra Pradesh and Telangana, Uzhavar Sandhai in Tamil Nadu, Shetkari Bazaars in Maharashtra and Raitha Santhe in Karnataka.

Direct marketing allows farmers to skip multiple layers in their transactions and benefits by skipping of intermediary margins. Though recommended in the Model APMC Act & Rules, very few of States have issued such licenses for direct sourcing.

Contract farming

Contract farming can address lack of market connectivity, long chain of market intermediaries, ignorance about the buyer demands, etc. Contract farming has a constraint from farmers' perspective, in that they are limited in their production to the direct demand and their growth is linked to the contractor's capacity to grow the market share. Though contract farming is not an entire solution for problems in agricultural marketing, it can very well be leveraged in certain regions and for specific crops for increasing farmers' income. {We have discussed this in current module}

Private wholesale markets

These have not yet developed in India and states need to liberalise the marketing regulations to promote development of private markets.

Organised retailing

Organised retailing on the lines of SAFAL (the fruit & vegetable marketing subsidiary of Mother Dairy) which is an example of indigenously organised retailing network. SAFAL operates in Delhi out of approximately 400 retail outlets, and sells about 350 tonnes of fresh produce daily in Delhi-NCR markets.

Farmer Producer Organisations (FPOs)

Organising producers into formal management practices help to take collective decisions on cultivation to make the best use of market intelligence, as well creates opportunities for producers to get involved in value adding decisions and activities such as input supply, credit, pre-conditioning, processing, marketing and distribution. The aggregation of farmer into FPOs (cooperatives/SHGs/FIGs/ Producer company), helps their integration into the supply chain.

Cooperatives in agricultural marketing

Cooperatives are organised to aggregate farmers for establishing scale in their production and marketing activities, besides easing access to credit and other services. Notable example is in the dairy sector – AMUL.

6

DEVELOPMENT OF AGRICULTURAL MARKETING

The term agricultural marketing is composed of two words- agriculture and marketing. Agriculture, generally means growing and/or raising of crops and livestock while marketing encompasses a series of activities involved in moving the goods from the point of production to point of consumption. Agricultural marketing essentially involves the buying and selling of agricultural produce.

According to the Indian Council of Agricultural Research, Agricultural marketing involves three important functions:

- Assembling or concentration of goods.
- Preparation for consumption (processing).
- Distribution of agricultural products.

According to the National Commission on Agriculture, Agricultural marketing is a process which starts with a decision to produce a saleable farm commodity, and it involves all aspects of market structure or system, both functional and institutional, based on technical and economic considerations and includes pre-harvest and post-harvest operations of –

- Assembling,
- Grading,
- Storage,
- Transportation, and
- Distribution.

Agricultural marketing plays an important role not only in stimulating production and consumption, but also in accelerating the pace of economic development. Its dynamic functions are of primary importance in promoting economic development. For this reason, it has been described as the most important multiplier of agricultural development.

An efficient marketing system is essential to maintain and accelerate the pace of increasing production through technological development, but for success in this programme, the farmers must receive remunerative prices for their produce. Otherwise, they would not be interested in increasing their production.

According to Thomsen – the study of agricultural marketing comprises all the operations, and the agencies conducting them, involved in the movement of farm produced foods, raw materials and their derivatives, such as textiles, from the farms to the final consumers, and the effects of such operations on farmers, middlemen and consumers.

According to the National Commission on Agriculture – Agricultural Marketing is a process which starts with a decision to produce a saleable farm commodity, and it involves all the aspects of market structure or system, both functional and institutional, based on technical and economic considerations, and include pre- and post-harvest operations viz., assembling, grading, storage, transportation and distribution.

Marketing of agricultural goods in India is of no mean significance. Agricultural marketing is one of the manifold problems, which has direct bearing upon the prosperity of the cultivator. Agricultural marketing, in its wide sense, comprises all the operations involved in the movement of goods and raw materials from the field to the final consumer.

It includes handling of product at the farm, initial processing, grading and packing in order to maintain and enhance quality and avoid wastage. Unfortunately, the present system of marketing of agricultural goods in India is extremely defective and needs a thorough overhauling.

The close inter-relationship between agricultural production (farming), agricultural finance (credit), and agricultural marketing (sale of farm products) has been recognised by the Government as well as by the experts on agricultural problems.

Farm products cover not only the products arising out of cultivation, but also products arising from dairy farming, poultry farming, fruit cultivation, vegetable farming, etc. In this wider sense, milk and milk products, eggs, fruits, vegetables, honey are also agricultural goods.

In spite of developed agricultural marketing with respect to highly sophisticated means of transport facilities, improved form of standardization and grading, advanced communication system, scientific method of storage and warehousing, the Indian agriculture is still for behind.

Therefore, it is required that wide and effective distribution channels should be used for different types of marketing, which exercise different functions in the marketing of agricultural goods. The small-scale production of farm products further leads to concentration, equalization and dispersion. As a result, these agricultural goods necessarily move through some important wholesale markets such as – jobbing market and secondary market.

MARKETING CHANNELS

I. Meaning: Farmers producing agricultural produce are scattered in remote villages while consumers are in semi-urban and urban areas. This produce has to reach consumers for its final use and consumption. There are different agencies and functionaries through which this produce passes and reaches the consumer. A market channel or channel of distribution is therefore defined as a path traced in the direct or indirect transfer of title of a product as it moves from a producer to an ultimate consumer or industrial user. Thus, a channel of distribution of a product is the route taken by the ownership of goods as they move from the producer to the consumer or industrial user.

II. Factors affecting channels: There are several channels of distribution depending upon type of produce or commodity. Each commodity group has slightly different channel. The factors are :

1. Perishable nature of produce. e.g. fruits, vegetables, flowers, milk, meat, etc.
2. Bulk and weight–cotton, fodders are bulky but light in weight.
3. Storage facilities.
4. Weak or strong marketing agency.
5. Distance between producer and consumer. Whether local market or distant market.

I. Types of Market Channels:

Some of the typical marketing channels for different product groups are given below:

A. Channels of rice:

1. Producer–miller->consumer (village sale)
2. Producer–miller->retailer–consumer (local sale)

3. Producer–wholesaler->miller–retailer–consumer
4. Producer–miller–cum-wholesaler-retailer-consumer
5. Producer–village merchant–miller–retailer–consumer
6. Producer–govt. procurement–miller–retailer–consumer

B. Channel of other foodgrains:

1. Producer - consumer (village sale)
2. Producer–village merchant–consumer (local sale)
3. Producer–wholesaler-cum-commission agent retailer–consumer
4. Producer–primary wholesaler–secondary wholesaler– retailer– Consumer
5. Producer–Primary wholesaler–miller–consumer (Bakers).
6. Producer->govt.procurement–retailer–consumer.
7. Producer–government–miller–retailer–consumer.

C. Channels of cotton:

1. Producer–village merchant–wholesaler or ginning factory– wholesaler in lint–textile mill (consumer)

2. Producer–Primary wholesaler–ginning factory–secondary wholesaler–consumer (Textile mill)

3. Producer– Trader– ginning factory– wholesaler in lint– consumer (Textile mill)

4. Producer–govt. agency–ginning factory–consumer (Textile mill).

5. Producer–Trader–ginning factory–wholesaler–retailer– consumer (non-textile use).

D. Channels of Vegetables:

1. Producers–consumer (village sale)
2. Producer–retailer–consumer (local sale)
3. Producer–Trader–commission agent–retailer–consumer.
4. Producer–commission agent–retailer–consumer
5. Producer–primary wholesaler–secondary wholesaler– retailer– consumer (distant market).

E. Channels of Fruits:

1. Producer–consumer (village sale)
2. Producer–Trader–consumer (local sale)

3. Producer–pre-harvest contractor–retailer–consumer

4. Producer–commission agent–retailer–consumer.

5. Producer–pre-harvest contractor–commission agent– retailer–consumer

6. Producer–commission agent–secondary wholesaler– retailer–consumer (distant market).

These channels have great influence on marketing costs such as transport, commission charges, etc. and market margins received by the intermediaries such as trader, commission agent, wholesaler and retailer. Finally this decides the price to be paid by the consumer and share of it received by the farmer producer. That channel is considered as good or efficient which makes the produce available to the consumer at the cheapest price also ensures the highest share to the producer.

AGRICULTURAL MARKETING DEVELOPMENT

Efforts to develop agricultural marketing have, particularly in developing countries, tended to concentrate on a number of areas, specifically infrastructure development; information provision; training of farmers and traders in marketing and post-harvest issues; and support the development of an appropriate policy environment. In the past, efforts were made to develop government-run marketing bodies but these have tended to become less prominent over the years.

Concept of Agricultural Marketing:

Agricultural marketing system is an efficient way by which the farmers can dispose their surplus produce at a fair and reasonable price. Improvement in the condition of farmers and their agriculture depends to a large extent on the elaborate arrangements of agricultural marketing.

The term agricultural marketing include all those activities which are mostly related to the procurement, grading, storing, transporting and selling of the agricultural produce. Thus Prof. Faruque has rightly observed: "Agricultural marketing comprises all operations involved in the movement of farm produce from the producer to the ultimate consumer. Thus, agricultural marketing includes the operations like collecting, grading, processing, preserving, transportation and financing."

Present State of Agricultural Marketing in India

In India four different systems of agricultural marketing are prevalent:

1. Sale in Villages:

The first method open to the farmers in India is to sell away their surplus produce to the village moneylenders and traders at a very low price. The moneylender and traders may buy independently or work as an agent of a bigger merchant of the nearly mandi. In India more than 50 per cent of the agricultural produce are sold in these village markets in the absence of organized markets.

2. Sale in Markets:

The second method of disposing surplus of the Indian farmers is to sell their produce in the weekly village markets popularly known as 'hat' or in annual fairs.

3. Sale in Mandis:

The third form of agricultural marketing in India is to sell the surplus produce though mandis located in various small and large towns. There are nearly 1700 mandis which are spread all over the country. As these mandis are located in a distant place, thus the farmers will have to carry their produce to the mandi and sell those produce to the wholesalers with the help of brokers or 'dalals'. These wholesalers of mahajans again sell those farm produce to the mills and factories and to the retailers who in turn sell these goods to the consumers directly in the retail markets.

4. Co-operative Marketing:

The fourth form of marketing is the co-operative marketing where marketing societies are formed by farmers to sell the output collectively to take the advantage of collective bargaining for obtaining a better price.

Defects of Agricultural Marketing in India

Following are some of the main defects of the agricultural marketing in India:

1. Lack of Storage Facility:

There is no proper storage or warehousing facilities for farmers in the villages where they can store their agriculture produce. Every year 15 to 30 per cent of the agricultural produce are damaged either by rats or rains due to the absence of proper storage facilities. Thus, the farmers are forced to sell their surplus produce just after harvests at a very low and un-remunerative price.

2. Distress Sale:

Most of the Indian farmers are very poor and thus have no capacity to wait for better price of his produce in the absence of proper credit facilities. Farmers often have to go for even distress sale of their output to the village moneylenders-cum-traders at a very poor price.

3. Lack of Transportation:

In the absence of proper road transportation facilities in the rural areas, Indian farmers cannot reach nearby mandis to sell their produce at a fair price. Thus, they prefer to sell their produce at the village markets itself.

4. Unfavourable Mandis:

The condition of the mandis are also not at all favourable to the farmers. In the mandis, the farmers have to wait for disposing their produce for which there is no storage facilities. Thus, the farmers will have to lake help of the middleman or dalal who lake away a major share of the profit, and finalizes the deal either in his favour or in favour of arhatiya or wholesalers. A study made by D.S. Sidhu revealed that the share of middlemen in case of rice was 31 per cent, in case of vegetable was 29.5 per cent and in case of fruits was 46.5 per cent.

5. Intermediaries:

A large number of intermediaries exist between the cultivator and the consumer. All these middlemen and dalals claim a good amount of margin and thus reduce the returns of the cultivators.

6. Unregulated Market's:

There are huge number of unregulated markets which adopt various malpractices. Prevalence of false weights and measures and lack of grading and standardization of products in village markets in India are always going against the interest of ignorant, small and poor farmers.

7. Lack of Market Intelligence:

There is absence of market intelligence or information system in India. Indian farmers are not aware of the ruling prices of their produce prevailing in big markets. Thus, they have to accept any un-remunerative price for their produce as offered by traders or middlemen.

8. Lack of Organisation:

There is lack of collective organisation on the part of Indian farmers. A very small amount of marketable surplus is being brought to the markets by a huge number of small farmers leading to a high transportation cost. Accordingly, the Royal Commission on Agriculture has rightly observed, "So long as the farmer does not learn the system of marketing himself or in cooperation with others, he can never bargain better with the buyers of his produce who are very shrewd and well informed."

9. Lack of Grading:

Indian farmers do not give importance to grading of their produce. They hesitate to separate the qualitatively good crops from bad crops. Therefore, they fail to fetch a good price of their quality product.

10. Lack of Institutional Finance:

In the absence of adequate institutional finance, Indian farmers have to come under the clutches of traders and moneylenders for taking loan. After harvest they have to sell their produce to those moneylenders at unfavourable terms.

11. Unfavourable Conditions:

Farmers are marketing their product under advice circumstances. A huge number of small and marginal farmers are forced by the rich farmers, traders and moneylenders to fall into their trap to go for distress sale of their produce by involving them into a vicious circle of indebtedness. All these worsen the income distribution pattern of the village economy of the country.

Remedial Measures for Improvement of Agricultural Marketing:

Improvement of the agricultural marketing in India is utmost need of the hour.

The following are some of the measures to be followed for improving the existing system of agricultural marketing in the country:

(i) Establishment of regulated markets.

(ii) Establishment of co-operative marketing societies.

(iii) Extension and construction of additional storage and warehousing facilities for agricultural produce of the farmers.

(iv) Expansion of market yards and other allied facilities for the new and existing markets.

(v) Provision is made for extending adequate amount of credit facilities to the farmers.

(vi) Timely supply of marketing information's to the farmers.

(vii) Improvement and extension of road and transportation facilities for connecting the villages with mandis.

(viii) Provision for standardisation and grading of the produce for ensuring good quality to the consumers and better prices for the farmers.

(ix) Formulating suitable agricultural price policy by the Government for making a provision for remunerative prices of agricultural produce of the country.

AGRICULTURAL MARKET INFRASTRUCTURE

Efficient marketing infrastructure such as wholesale, retail and assembly markets and storage facilities is essential for cost-effective marketing, to minimize post-harvest losses and to reduce health risks. Markets play an important role in rural development, income generation, food security, and developing rural-market linkages. Experience shows that planners need to be aware of how to design markets that meet a community's social and economic needs and how to choose a suitable site for a new market. In many cases sites are chosen that are inappropriate and result in under-use or even no use of the infrastructure constructed. It is also not sufficient just to build a market: attention needs to be paid to how that market will be managed, operated and maintained. Most market improvements that have been only aimed at infrastructure upgrading and have not guaranteed maintenance and management have failed within a few years

Rural assembly markets are located in production areas and primarily serve as places where farmers can meet with traders to sell their products. These may be occasional (perhaps weekly) markets, such as haat bazaars in India and Nepal, or permanent. Terminal wholesale markets are located in major metropolitan areas, where produce is finally channelled to consumers through trade between wholesalers and retailers, caterers, etc. The characteristics of wholesale markets have changed considerably as retailing changes in response to urban growth, the increasing role of supermarkets and increased consumer spending capacity. These changes may require responses in the way in which traditional wholesale markets are organized and managed.

Retail marketing systems in western countries have broadly evolved from traditional street markets through to the modern hypermarket or out-of-town shopping center. In developing countries, there remains scope to improve agricultural marketing by constructing new retail markets, despite the growth of supermarkets, although municipalities often view markets primarily as sources of revenue rather than infrastructure requiring development. Effective regulation of markets is essential. Inside a market, both hygiene rules and revenue collection activities have to be enforced. Of equal importance, however, is the maintenance of order outside the market. Licensed traders in a market will not be willing to cooperate in raising standards if they face competition from unlicensed operators outside who do not pay any of the costs involved in providing a proper service

MARKET INFORMATION

Efficient market information can be shown to have positive benefits for farmers and traders. Up-to-date information on prices and other market factors enables farmers to negotiate with traders and also facilitates spatial distribution of products from rural areas to towns and between markets. Most governments in developing countries have tried to provide market information services to farmers, but these have tended to experience problems of sustainability. Moreover, even when they function, the service provided is often insufficient to allow commercial decisions to be made because of time lags between data collection and dissemination. Modern communications technologies open up the possibility for market information services to improve information delivery through SMS on cell phones and the rapid growth of FM radio stations in many developing countries offers the possibility of more localised information services. In the longer run, the internet may become an effective way of delivering information to farmers.

However, problems associated with the cost and accuracy of data collection still remain to be addressed. Even when they have access to market information, farmers often require assistance in interpreting that information. For example, the market price quoted on the radio may refer to a wholesale selling price and farmers may have difficulty in translating this into a realistic price at their local assembly market. Various attempts have been made in developing countries to introduce commercial market information services but these have largely been targeted at traders, commercial farmers or exporters. It is not easy to see how small, poor farmers can generate sufficient income for a commercial service to be profitable although in India a new service introduced by Thomson Reuters was reportedly used by over 100,000 farmers in its first year of operation. Esoko in West Africa attempts to subsidize the cost of such services to farmers by charging access to a more advanced feature set of mobile-based tools to businesses.

MARKETING TRAINING

Farmers frequently consider marketing as being their major problem. However, while they are able to identify such problems as poor prices, lack of transport and high post-harvest losses, they are often poorly equipped to identify potential solutions. Successful marketing requires learning new skills, new techniques and new ways of obtaining information. Extension officers working with ministries of agriculture or NGOs are often well-trained in horticultural production techniques but usually lack knowledge of marketing or post-harvest handling.

Agricultural marketing support

Most governments have at some stage made efforts to promote agricultural marketing improvements. In the United States the Agricultural Marketing Service (AMS) is a division of USDA and has programs that provide testing, support standardization and grading and offer market news services. AMS oversees marketing agreements and orders research and promotion programs. It also purchases commodities for federal food programs. USDA also provides support to agricultural marketing work at various universities. In the United Kingdom, support for marketing of some commodities was provided before and after the Second World War by boards such as the Milk Marketing Board and the Egg Marketing Board. These boards were closed down in the 1970s. As a colonial power, Britain established marketing boards in many countries, particularly in Africa. Some continue to exist although many were closed at the time of the introduction of structural adjustment measures in the 1990s.

Several developing countries have established government-sponsored marketing or agribusiness units. South Africa, for example, started the National Agricultural Marketing Council (NAMC) as a response to the deregulation of the agriculture industry and closure of marketing boards in the country. India has the long-established National Institute of Agricultural Marketing. These are primarily research and policy organizations, but other agencies provide facilitating services for marketing channels, such as the provision of infrastructure, market information and documentation support. Examples from the Caribbean include the National Agricultural Marketing Development Corporation (NAMDEVCO) in Trinidad and Tobago and the New Guyana Marketing Corporation in Guyana.

Agricultural value chain

The agricultural value chain concept has been used since the beginning of the millennium, primarily by those working in agricultural development in developing countries. Although there is no universally accepted definition of the term, it normally refers to the whole range of goods and services necessary for an agricultural product to move from the farm to the final customer or consumer.

Agricultural value chain finance

Agricultural value chain finance is concerned with the flows of funds to and within a value chain to meet the needs of chain actors for finance, to secure sales, to buy inputs or produce, or to improve efficiency. Examining the potential for value chain finance involves a holistic approach to analyze the chain, those working in it, and their inter-linkages. These linkages allow financing to flow through the chain. For example, inputs can be provided to farmers and the cost can be repaid directly when the product is delivered, without need for farmers taking a loan from a bank or similar institution. This is common under contract farming arrangements. Types of value chain finance include product financing through trader and input supplier credit or credit supplied by a marketing company or a lead firm. Other trade finance instruments include receivables financing where the bank advances funds against an assignment of future receivables from the buyer, and factoring in which a business sells its accounts receivable at a discount. Also falling under value chain finance are asset collateralization, such as on the basis of warehouse receipts, and risk mitigation, such as forward contracting, futures and insurance

Marketing systems: Functions, agents, enterprises and channels

All marketing systems have evolved within the constraints and conditions placed upon them by the production sector and by the nature of the goods being marketed. The type of product, the number, size and density of producers, the infrastructure and the policy and institutional environments all determine the type of marketing system and the effectiveness with which it operates.

Important points · Marketing includes all business activities involved in the flow of goods and services from producers to consumers.· For a consumer or producer, the objectives of marketing are to maximise benefits from the available resources and to expand marketing operations to increase wealth.· From a societal viewpoint, the major objectives of marketing are to:- assist in the efficient allocation of resources- create wealth and promote economic growth- improve income distribution among different sectors of the economy- maintain stability of supply and demand for marketed goods.· The main characteristics of agricultural and livestock marketing in Africa are:- Marketing begins at the level of individual farmers.- Producers are widely dispersed.- Agricultural and livestock commodities are perishable and seasonal in supply.- Institutional supports are limited.· Marketing and trade allow specialisation of activities. This leads to enhanced resource-useefficiency and economic growth. With economic development, the tasks and activities of marketing further increase, creating employment and other avenues for development.· Some important prerequisites to market development are:- proper linkages between rural areas and urban centres of consumption- a conducive policy institutional environment for marketers to operate effectively and for markets to expand beyond the basic need levels of consumers and producers- a dynamic relationship between supply and demand.· The marketing system and its structure influence the determination of a commodity's market price.· The six main elements of price theory used to explain market prices are:- perfect competition- imperfect competition- horizontal and vertical integrations of markets- separation of markets- product differentiation- seasonal and cyclic variations.

A marketing system is comprised of a number of elements: the particular products (e.g. butter only, or butter and raw milk) and their characteristics being transferred from producer to consumer; the characteristics of participants (e.g. the producer, the trader, the consumer); the functions or roles that each participant performs in the market; and the locations, stages, timetable and physical infrastructures involved.

When we talk of describing, quantifying or analysing a particular marketing system, there is an implicit assumption that we can distinguish the elements of that system from other economic activities. Analyses of marketing systems usually include a quantification of the flows and of the value added, costs and profit margins at each stage in the system.

Marketing functions:

Marketing is sometimes thought of as simply the process of buying and selling. Its tasks are much more extensive than this simple description. For a marketing system to be operative and effective, there are three general types of functions which it must provide.

- Exchange functions:
- Buying
- Selling
- Pricing.
- Physical functions:
- Assembling
- Transport and handling
- Storage
- Processing and packaging
- Grading and standardisation.
- Facilitating functions:
- Financing and risk-bearing
- Market information
- Demand and supply creation
- Market research.

Exchange functions are what is commonly thought of as marketing. They involve finding a buyer or a seller, negotiating price and transferring ownership (but not necessarily physical transfer). These functions take place at the "market" - that is, the physical meeting point for buyers and sellers at the point of production or via some other means of communication. At this point, formal or informal property rights are important to ensure the reliable transfer of ownership and to guarantee legality (e.g. that animals on sale were not stolen and will not be reclaimed).

Physical functions enable the actual flow of commodities through space and time from producer to consumer and their transformation to a form desirable to the consumer. Assemblying or concentrating the product at convenient points allows its economical transport (i.e. getting enough animals together to transport cheaply). This is a valuable function which is often overlooked in the public perception of traders. Storage allows the commodity to be held until peak season demand, thereby stabilising supply. Processing transforms the commodity into the products desired by the consumers. Grading and standardisation allow the consumer to be more confident of the characteristics of the good being purchased.

Financing and risk-bearing are two important facilitating functions. The owner of goods at any marketing stage must sacrifice the opportunity to use the working capital needed to buy those goods elsewhere. Or the owner must borrow that capital. In either case, capital must be provided by the trader or by some lending source. Regardless, cost is involved. Further, there is an implicit cost in the risk of losing all or part of that capital through theft, spoilage, mortality or changing market conditions. Without the willingness to provide the capital and to bear these costs, no stage of the market chain could function. Other facilitating functions enable producers to respond to consumer needs and thus provide goods in the locations, quantity and form desired.

These functions create the marketing environment, whose elements are:

Market and facilities - including all of the physical infrastructure that a market may depend on.

Market information and intelligence - including informal and formal communication systems, and standard weights and grades on which market information depends.

Institutional environment - including the government policy environment, regulations and supporting legislation.

Marketing Agents

Market actors are defined the roles they play, by the nature of their financing activities and their responsibility to ownership (whether to a centralised public office, to a private purchaser of the marketing: service or self-responsibility as in the case of independent private enterprises). Actors in the market can choose between specialising in one activity or integrating a number of activities into one enterprise in a vertical or horizontal manner. A specialised enterprise can offer its customers more individual attention and provide the exact quality and form of goods desired (e.g. local butcher). Thus, the roles of actors are often difficult to separate. The roles of vertically-integrated actors are likely to overlap with those of more specialised agents in the market.

Country buyers often carry out the initial task of assembling goods from dispersed farms or local rural markets. These buyers may be farmers, shopkeepers, itinerant traders or some co-operative or government-buying agency.

The role of wholesalers is to transfer goods from producers or country buyers to retailers or other wholesalers. Thus, their role may overlap with that of country buyers, in that they may deal directly with producers. They often finance the movement of goods themselves and consequently bear the cost of marketing risks. In the African livestock trade, there is a tendency for there to be a number of stages in the wholesale trade, as animals are assembled into larger and larger herds for subsequent trekking to urban centres. To operate profitably, wholesalers must be especially well-informed about current market prices and conditions, since the costs of market risks increase with the number of stock being handled.

Commission agents may sometimes operate on behalf of wholesalers for a percentage of the price paid. Although they act in the same way as wholesalers, the risk remains with the owner of the goods. Brokers offering an intimate knowledge of the market act to bring buyers and sellers together. In West Africa, livestock brokers also serve to enforce informal market rules by monitoring transactions, assuring the integrity of each party in the transaction and guaranteeing the negotiated price will be paid. Thus, they contribute in several ways to the exchange functions by facilitating buying and selling, and reinforcing the informal system which enforces contracts.

Processors transform the good either partially or completely into the form to be consumed. In the African livestock trade, processing is often carried out on a large scale by government agencies who also operate as wholesalers. They may also sell their processing services to smaller traders.

Retailers present the good to the consumer in the manner, location and form desired. In the case of livestock, they may also carry out processing activities (e.g. butchering).

Marketing enterprises and channels

Enterprises of four types normally fulfill the roles of middlemen described above. These are:

- Independent, locally-based private enterprises
- Co-operatives
- Marketing boards and other state enterprises
- Transnational companies.

Independent, locally-based private enterprises operate with capital owned directly by the operators and their partners, or in some cases by shareholders. Although not always large in scale of operation, these make up the greatest number of agriculture and livestock enterprises. Great variety exists in their level and degree of sophistication. Sometimes foreign-owned operations may' occupy important roles in this niche, particularly in foreign trade of livestock products.

Co-operatives have the potential to improve marketing efficiency. They can reduce marketing costs. For example, a village livestock marketing co-operative could co-ordinate the production schedules of small farmers, so that sufficient animals would reach market age at the same time, allowing truck transport to markets and lowering per unit transport costs. Co-operatives can also be used to counteract imperfect competition (monopsony/oligopoly power) among buyers, by creating greater bargaining power among producers. Typically they are used to distribute credit or subsidised inputs. In Africa they have been more successful when they have confined themselves at first to one' simple function which is important to all members, attempting only later to expand their role. In order to be successful in the long run, a co-operative must be able to carry out marketing functions with lower cost or effort than available alternatives. If this ability is not perceived by members, co-operatives are likely to break down. Since the ownership of co-operatives, by definition, lies in the hands of those who use its services (and who are thus entitled to any profits), a distinction must be made between farmer-owned and -controlled co-operatives and parastatals. Parastatals are co-operatives in name only, since they are government controlled. They may serve as taxation mechanisms or to promote government support. Private co-operatives are likely to be more efficient than parastatals, because of ownership incentives. Some co-operatives are difficult to classify, such as the Kenya Co-operatives Creamery (KCC). Although nominally a private cooperative, the KCC acts as a parastatal because of government-sanctioned monopoly and regulatory powers.

Marketing boards and other state enterprises, although popular with many African governments, have been much criticised. They are set up by government direction with government capital. Major operating decisions are subject to approval by the responsible minister. Parastatals are slightly more independent. Although government financed, they are autonomous in terms of handling funds, recruiting staff and making operational decisions.

The objectives of establishing such public intermediaries are: to raise the bargaining power of agricultural producers via an imposed monopoly on sales; 'to set up needed market and processing facilities; to raise the scale of operation and thus to capture economies of scale; and to stabilise market supply and prices. They often fail to achieve these objectives because of inappropriate policies, poor management and lack of knowledge. Attempts to replace private markets usually fail because the detailed information necessary to operate may be too dispersed to gather. Managers succumb to patronage and corruption, and incentives for efficient operation are usually lacking. The Kenya Meat Commission (KMC) was, until recently, a parastatal set up to buy and process cattle and to market the products. Although potential economies of scale existed, these were not achieved because capacity was under-utilised and per unit costs were higher. Slaughterhouses built to handle peak seasonal supply are usually under-utilised during other seasons. Parastatals with a mandate to buy at fixed prices from all producers also suffer from high costs of cattle purchases in pastoral areas, where such sales are widely dispersed.

Transnational companies often succeed because of their access to processing technology and external markets. By definition, they operate in countries other than that of their headquarters. They can assist market development by facilitating the movement of skills and capital to areas where they are in short supply, potentially contributing to the levelling of commercial expertise.

When considering the relative advantages of each of these enterprises, attention must be given to the particular environment of livestock marketing in Africa. Its marketing structures are more complicated and differentiated than those in a developed country where production is much more specialised. Further, issues of equity and income distribution between producers are more acute and must be considered in the policy decision to promote certain types of enterprise.

All of the goods in a particular market are unlikely to pass through the same set of agents. Usually goods pass through a variety of market channels as a result of varying degrees of vertical integration existing in the same market. Figure 5.4 illustrates a marketing system showing multiple marketing channels. At times, some intermediaries are bypassed, while in others, goods pass through a large number of hands. Mote that at any one level of the market, such as at level A, the sum of the percentage flows in the diagram is always 100. This type of diagram can be helpful in basic planning for new investment in marketing, by identifying both the channels where volume is highest and other channels which could be further developed.

AGRICULTURAL AND FOOD MARKETING

As individuals within a society become more specialised in their economic activities, they come to rely upon others to supply at least some of the products and services which they need. Thus begins a process of exchange between buyers and sellers. For a while buyers and sellers remain in immediate contact and each party is able to determine what the other needs and values and, therefore, will be willing to exchange. As the economy develops the number and types of exchanges expand, there is a concomitant need for increasingly specialised marketing services such as physical distribution, storage, grading, market information gathering and so. The number of participants also increases with many of the specialised services being provided by intermediaries between the seller and ultimate buyer. Few buyers and sellers are in direct contact with one another and communication between them is channelled through a complex marketing system. This introductory chapter is devoted to exploring the nature of marketing and marketing systems.

The importance of agricultural and food marketing

In many countries, and virtually every less developed country (LDC), agriculture is the biggest single industry. Agriculture typically employs over fifty percent of the labour force in LDCs with industry and commerce dependent upon it as a source of raw materials and as a market for manufactured goods. Hence many argue that the development of agriculture and the marketing systems which impinge upon it are at the heart of the economic growth process in LDCs. Moreover as Kriesberg points out; in LDCs the consumer frequently spends in excess of fifty percent of the household's income on basic foodstuffs - much of which is inadequate both in quality and nutritional content. By contrast Americans spend approximately twelve percent of their total disposable income on food. In Western Europe the figure ranges from about sixteen to nineteen percent of disposable income. Furthermore, whereas in developed countries the poor are relatively few in number, and therefore it is economically possible to establish special food distribution programmes to meet their needs, the scale of poverty in most LDCs is such that the commercial marketing system must be relied upon to perform the task of food distribution to poor and not-so-poor alike. This being so, it is imperative that the marketing system performs efficiently.

Economic development itself provides the impulse towards more sophisticated and more efficient marketing systems. Dixie suggests that as countries experience economic growth, their rate of urbanisation tends to increase substantially. Whereas the rate of population growth, in developing countries, averages around three percent per annum, their cities and towns are increasing their populations at about four percent per annum. In essence, this means that the number of people, in urban areas, needing to be fed by rural people, will double within sixteen years. This has clear implications for agricultural production and the marketing systems that direct that production and distribute the output to the points of its consumption. Subsistence farming is likely to diminish in importance as farmers respond to the increased opportunities that development and urbanisation create; farms are likely to decrease in number whilst increasing in size; and agriculture will probably become less labour intensive and more capital intensive.

Dixie also highlights the potential contribution of agricultural and food marketing, towards attempts to improve rural incomes in developing countries. The inequality of incomes between the rural and urban areas draws people away from agricultural production and places great stress upon the infrastructure and social services of a country's towns and cities. Nowhere was this more dramatically demonstrated than in Nigeria when petroleum oil was discovered and then exploited in the 1970s. A large number of jobs were created in the urban areas and people abandoned agricultural production in large numbers. Nigeria became a net importer of many agricultural products of which it had formerly been a net exporter. For as long as the world price for petroleum remained high the economy thrived and could well afford the food import bill. However, as soon as the world price for oil fell, the food import bill became a serious burden. Nigeria would only have avoided this scenario if it had been able to motivate people to continue in agriculture and this would only have been possible if the disparity between urban and rural incomes had been reduced. Rurally based enterprises, including small-holdings, can greatly improve their earning potential by adopting a market orientation. They can be encouraged to add value to commodities by adding to their utility. Value added products normally carry a higher margin than raw commodities.

Another development which has in recent times increased interest in marketing practies is the trend, in many developing countries, towards market liberalisation as part of economic structural adjustment programmes (ESAPs). The view that direct and indirect government participation in production and distribution had brought about structural distortions in economies has become widely accepted. Measures intended to correct these distortions include a return to market prices for all products and resources, the encouragement of a competitive private sector and the commercialisation, and sometimes privatisation, of all or some of the functions of marketing parastatals. All of this requires a better understanding of marketing practices and processes within the country implementing ESAPs, in general, and within the agricultural marketing parastatals affected, in particular.

So far this discussion has been set in the context of commercial marketing but social marketing should also be acknowledged. Social marketing identifies human needs in non-competitive economies and/or sectors of society and defines the means of delivering products and services to meet these needs. The marketing mix of social marketing strategies is evaluated using quite different criteria from those employed in assessing purely commercial marketing strategies. Criteria such as the percentage of the target population reached with the technology, products, processes or services, quantities produced and distributed and uptake of the product, service or technology are more often employed. Benefits are measured in terms of development goals, such as improved nutritional status or increased rural incomes. The use of economic criteria is usually limited to the latter and to selecting the least-cost strategy to achieve a quantitative goal. However, the criteria used to evaluate commercial marketing strategies should not automatically be eliminated, because these improve the efficiency of some aspects of social marketing strategy without preventing the attainment of social objectives.

The marketing concept and marketing systems

Marketing is not simply an extension of the production process but its only purpose as Adam Smith emphasised when, in his text The Wealth of Nations (1776), he said that:

"Consumption is the sole end purpose of all production: and the interest of the producer ought to be attended to only so far as it may be necessary for promoting that of the consumer."

Dixie relates what he describes as a definition of marketing which is

"The series of services involved in moving a product (or commodity) from the point of production to the point of consumption."

This is a definition which many organisations, and governments, would recognise as describing their own activities in commodity marketing. Indeed in many developing countries it aptly describes, or in some cases, did in the past describe, the functions carried out by marketing parastatals with respect to staple foods. However, as Dixie himself points out, the definition omits two key elements of any definition of marketing production to effuse the marketing concept, i.e. a customer orientation and inbuilt sustainability. Gaedeke and Tootelian offer an alternative definition which overcomes the problems caused by these two omissions:

"... a management orientation focusing all the activities of the organization on satisfying customer needs and wants, thereby helping achieve the organization's long-range objectives."

This definition promotes a customer orientation and since the organisation's long-term objectives will include it's own continued existence it takes account of the need for sustainability. Moreover, this definition of the marketing concept does not preclude non-profit making organisations. Marketing is just as relevant to development projects, aid agencies, extension service organisations, and the like, as it is to commercial enterprises. Thus the marketing concept is that an organisation achieves its goals through the provision of customer satisfaction. Put another way, marketing is the integrative force that matches production to customer needs and satisfaction. Marketing is not an activity to which an organisation turns its attention at the end of the production phase of operations. Rather marketing needs to be directing production in accordance with clear signals from the marketplace as what is needed by customers.

The marketing concept must be adopted throughout not only the entire organisation, but the entire marketing system. A system is a complex of interrelated component parts or sub-systems which have a defined common goal. Thus, an agricultural and food marketing system comprises all of the functions, and agencies who perform those activities, that are necessary in order to profitably exploit opportunities in the marketplace. Each of the components, or sub-systems, are independent of one another but a change in any one of them impacts on the others as well as upon the system as a whole.

There is a danger that the marketing concept will be adopted by some parts of the system but not others. Thus, for example, a food manufacturer may be trying hard to implement the marketing concept and offer products that meet the precise needs of a target market. If, however, the manufacturer has to rely upon a farming community that is still very much production oriented, for raw material supplies, then the overall marketing objectives may be frustrated. In the same way, if only some functions are performed according to the marketing concept then the system as a whole may not achieve a market orientation. For instance, the marketing department may set out to serve the market for a high quality fruits and vegetables, for which it can obtain premium prices, but if transportation is performed using the same open-topped bulk carrying wagons used to ship grain and other aggregates then it is unlikely that the enterprise will deliver the product in the right condition for the target market.

Marketing Sub-systems:

Rosson conceives of agricultural and food marketing systems as consisting of 4 main sub-systems; production, distribution, consumption and regulatory.

The key players in the chain of activities that connect food and agriculture are the farmer, (or other 'producers' such as fishermen), intermediaries, the food processors, and the consumer. In practice they each see the agricultural/ food marketing system from a perspective of self-interest and these interests are sometimes in conflict. Illustrative examples of some of the conflicts which typically arise are given in table 1.1.

TABLE 1.1 CONFLICT OF INTEREST IN AGRICULTURAL/FOOD MARKETING SYSTEMS

Key Players	Interests
Farmers	Maximum price, unlimited quantities
Manufacturers	Low purchase price, high quality
Traders and retailers	Low purchase price, high quality
Consumers	Low purchase price, high quality

The farmer's interest is focused on getting the best return from his produce, which usually equates to maximum price for unlimited quantities. Manufacturers want least cost, best quality produce from the farmer so that he can sell it at competitive, but profitable, prices. Traders and retailers want high quality and reliable supplies from the manufacturer or farmer, at the most competitive prices. Consumers are interested in obtaining high quality products at low prices. Clearly, there are conflicting interests here.

Marketing functions

A little earlier it was said that a marketing system has two distinct dimensions. One of those dimensions is the institutions, organisations and enterprises which participate in a market and the second is the functions that those participants perform. Kohls and Uhl have classified the functions involved in agricultural and food marketing processes as under three sets of functions of a marketing system

A. Exchange Functions	1. Buying
B. Physical Functions	2. Selling
C. Facilitating Functions	3. Storage
	4. Transportation
	5. Processing
	6. Standardisation
	7. Financing
	8. Risk Bearing
	9. Market Intelligence

Each of these functions add value to the product and they require inputs, so they incur costs. As long as the value added to the product is positive, most firms or entrepreneurs will find it profitable to compete to supply the service.

Exchange functions

Buying: The marketing concept holds that the needs of the customer are of paramount importance. A producer can be said to have adopted a market orientation when production is purposely planned to meet specific demands or market opportunities. Thus a contract farmer who wishes to meet the needs of a food processor manufacturing sorghum-based malted drinks will only purchase improved sorghum seed. He/she will avoid any inputs likely to adversely affect the storage and/or processing properties of the sorghum and will continually seek new and better inputs which will add further value to his/her product in the eyes of the customer. In making his/her buying decisions his underlaying consideration will be the effect upon the attractiveness of his/her output to the markets he/she is seeking to serve.

The buyer's motive is the opportunity to maintain or even increase profits and not necessarily to provide, for example, the best quality. Improving quality inevitably increases the associated costs. In some cases the market is insensitive to improvements in quality, beyond some threshold level, does not earn a premium price. Under such circumstances, the grower who perseveres and produces a 'better product', is not market oriented since he/she is ignoring the real needs of the consumer. The most successful agribusiness is the one which yields the largest difference between prices obtained and costs incurred.

Selling: Of the nine functions listed, this is probably the one which people find least difficulty in associating with marketing. Indeed to many the terms marketing and selling are synonymous. Kotler suggests that:

"Most firms practice the selling concept when they have over capacity. Their immediate aim is to sell what they can make rather than to make what they can sell."

There is no denying that 'high pressure selling' is practiced, where the interests of the consumer are far from foremost in the mind of the seller. This is not marketing. Enterprises adopt the marketing philosophy as a result of becoming aware that their own long term objectives can only be realised by consistently providing customer satisfaction. Whereas selling might create a consumer, marketing is about creating a customer. The difference is that marketing is about establishing and maintaining long term relationships with customers.

Selling is part of marketing in the same way that promotion, advertising and merchandising are components, or sub-components of the marketing mix. These all directed towards persuasion and are collectively known as marketing communications; one of the four elements of the marketing mix.

Physical functions

Storage: An inherent characteristic of agricultural production is that it is seasonal whilst demand is generally continous throughout the year. Hence the need for storage to allow a smooth, and as far as possible, uninterrupted flow of product into the market. Because he is dealing with a biological product the grower does not enjoy the same flexibility as his manufacturing counterpart in being able to adjust the timing of supply to match demand. It would be an exaggeration to suggest that a manufacturer can turn production on and off to meet demand - they too have their constraints- but they have more alternatives than does the agricultural producer. A manufacturer can, for example, work overtime, sub-contract work, and over a longer time horizon, the manufacturer can increase or decrease productive capacity to match the strength of demand.

In agriculture, and especially in LDCs, supply often exceeds demand in the immediate post-harvest period. The glut reduces producer prices and wastage rates can be extremely high. For much of the reminder of the period before the next harvest, the product can be in short supply with traders and consumers having to pay premium prices to secure whatever scarce supplies are to be had. The storage function is one of balancing supply and demand.

Both growers and consumers gain from a marketing system that can make produce available when it is needed. A farmer, merchant, co-operative, marketing board or retailer who stores a product provides a service. That service costs money and there are risks in the form of wastage and slumps in market demand, prices, so the provider of storage is entitled to a reward in the form of profit.

Transportation: The transport function is chiefly one of making the product available where it is needed, without adding unreasonably to the overall cost of the produce. Adequate performance of this function requires consideration of alternative routes and types of transportation, with a view to achieving timeliness, maintaining produce quality and minimising shipping costs.

Effective transport management is critical to efficient marketing. Whether operating a single vehicle or a fleet of vehicles, transportation has to be carefully managed, including cost monitoring - operations on different road types, fuel and lubrication consumption and scheduled and remedial maintenance and repair. Skillful management of all aspects of vehicle operations can also make a substantial contribution to efficient marketing especially with respect to optimum routing, scheduling and loading and off-loading; maximisation of shift hours available, maintaining the vehicle fleet at an optimum size, taking account of time constraints on delivery, and collection times and judicious management of vehicle replacement and depreciation. Transport managers also have to weigh the advantages and disadvantages of owning, hiring or leasing transport.

Processing: Most agricultural produce is not in a form suitable for direct delivery to the consumer when it is first harvested. Rather it needs to be changed in some way before it can be used. Kohls and Uhl observe that:

"The processing function is sometimes not included in a list of marketing functions because it is essentially a form changing activity."

However, it is for this very reason that processing ought to be included as a marketing function. The form changing activity is one of that adds value to the product. Changing green coffee beans into roasted beans, cassava into gari or livestock feed, full fruit bunches into palm oil or sugarcane into gur increases the value of the product because the converted product has greater utility to the buyer. How the form of produce is to be changed and the method to be used in bringing about such changes are marketing decisions. For example, some years ago when Ethiopia was looking to expand its tea business, a prototype manufacturing plant was established. The plant was capable of curing the tea and packing it in individual tea bags. At that point, tests were undertaken in which the product was compared with others already on the market. The results were encouraging. However, in the course of the marketing research, it was also discovered that ninety percent of the black tea consumed is blended and not the pure variety placed in tea bags by the Ethiopians. By going past the point of changing green leaf into high quality black tea, the Ethiopians were entering a nice market which is not what they intended at all. Timely marketing research would have directed Ethiopia to stop the form changing activity short of bagging since, at that time, Ethiopia did not have the acreage of tea, nor the resources, to develop a tea blending facility of its own. In the same way, a producer of fresh fruits may have pulping and/or canning facilities but if potential buyers want the flexibility of using the fruits in a variety of ways, then these stages of processing serve to reduce utility and value, rather than increasing them.

Of course, processing is not the only way of adding value to a product. Storing products until such times as they are needed adds utility and therefore adds value. Similarly, transporting commodities to purchasing points convenient to the consumer adds value. In short, any action which increases the utility of the good or service to prospective buyers also adds value to that product or service.

Facilitating functions

The facilitating functions include product standardisation, financing, risk bearing and market intelligence. Facilitating functions are those activities which enable the exchange process to take place. Marketing, in simple terms, is the act of supplying products to someone in exchange for something perceived to be of equal or greater value, (usually, but not always, a given sum of money). Facilitating functions are not a direct part of either the exchange of title or the physical movement of produce.

Standardisation: Standardisation is concerned with the establishment and maintenance of uniform measurements of produce quality and/or quantity. This function simplifies buying and selling as well as reducing marketing costs by enabling buyers to specify precisely what they want and suppliers to communicate what they are able and willing to supply with respect to both quantity and quality of product. In the absence of standard weights and measures trade either becomes more expensive to conduct or impossible altogether. In Nepal such was the diversity of weights and measures used with respect to grain within the country, that it was easier for some districts to conduct trade with neighbouring states in India than it was to do business with other districts within Nepal. Among the most notable advantages of uniform standards, are:

- Price quotations are more meaningful
- The sale of commodities by sample or description becomes possible
- Small lots of commodities, produced by a large number of small producers, can be assembled into economic loads if these supplies are similar in grade or quality

- Faced with a range of graded produce the buyer is able to choose the quality of product he/she is able and willing to purchase.

Quality differences in agricultural products arises for several reasons. Quality differences may be due to production methods and/or because of the quality of inputs used. Technological innovation can also give rise to quality differences. In addition, a buyer's assessment of a product's quality is often an expression of personal preference. Thus, for example, in some markets a small banana is judged to be in some sense 'better' than a large banana; white sugar is considered 'superior' to yellow sugar; long stemmed carnations are of 'higher quality' than short stemmed carnations; and white maize is 'easier to digest' than yellow maize. It matters not whether the criteria used in making such assessments are objective or subjective since they have the same effect in the marketplace. What does matter in marketing is to understand how the buyer assesses 'quality'.

Financing: In almost any production system there are inevitable lags between investing in the necessary raw materials (e.g. machinery, seeds, fertilizers, packaging, flavourings, stocks etc.) and receiving the payment for the sale of produce. During these lag periods some individual or institution must finance the investment. The question of where the funding of the investment is to come from, at all points between production and consumption, is one that marketing must address. Consider the problem of a food manufacturer who wishes to launch a range of chilled products in a developing country where few retail outlets have the necessary refrigeration equipment. This is a marketing problem. It might be solved by the food manufacturer buying refrigerators and leading these to retailers (or arriving a hire-purchase arrangement with retailers).

A common marketing problem, in developing countries, is the low level of incomes leading to low levels of effective demand for many products. The challenge to marketing is to somehow channel what income is available into effective demand. In the case of agricultural equipment marketing this might involve offering hire-purchase schemes where the prospective buyer makes payment in regular installments. During this time he/she is deemed to have hired the machine. If payments are not forthcoming, the machine can be recovered since its ownership remains with the seller up until the final payment is made, at which point the farmer is considered to have purchased the machine. Alternatively, the seller might set up leasing, rather than purchasing schemes where again the farmer is making regular payments but never takes title to the machine. Where a food item is being marketed, to a low income market, the seller can consider reducing the until price of the product by making the pack or lot size smaller. Another tactic is to make the product more affordable by using cheaper ingredients and/or packaging. Instant coffee can be sold at lower prices by substituting some of the coffee with chicory; the price of meat products is reduced by increasing the percentage of cereals in these products and including less meat and/or making use of less expensive parts of the animal such as entrails, offal, feet and head.

Marketing is also concerned with the financing of the enterprise itself. Here again some creative solutions can be developed. Where internal financing is insufficient for the purposes in view, an enterprise in a developing country can look to several alternatives including:

- development banks
- commercial banks
- shares issues
- credit co-operatives and/or credit unions

Where these sources of finance are considered inappropriate, or are simply not available to a particular enterprise, a strategic alliance in the form of a joint venture could be the answer. These are partnerships formed to exploit market opportunities more effectively and/or efficiently than either party can on its own. An enterprise, in a developing country, may engage in a joint venture with either an indigenous partner and/or with a foreign partner. The agreement between parties to a joint venture normally specifies their respective contributions of resources, share of management control, profit and risk.

Whatever the source of finance under consideration marketing has a role to play in evaluating the appropriateness of that source as well as identifying it in the first place. A common requirement is that marketing proposals include a forecast of the payback period. Those responsible for developing these proposals are best placed to evaluate the compatibility between the market opportunity under consideration and the alternative modes of financing it. Of specific interest is the prospect of the investment payback period matching the repayment schedule. Enterprises which finance long term investments through short term sources of finance are either badly misinformed or have adopted a high risk strategy.

Risk bearing: In both the production and marketing of produce the possibility of incurring losses is always present. Physical risks include the distruction or deterioration of the produce through fire, excessive heat or cold, pests, floods, earthquakes etc. Market risks are those of adverse changes in the value of the produce between the processes of production and consumption. A change in consumer tastes can reduce the attractiveness of the produce and is, therefore, also a risk. All of these risks are borne by those organisations, companies and individuals.

Risk bearing is often a little understood aspect of marketing. For example, when making judgements as to whether a particular price is a 'fair price' the usual reference point is the producer or supplier's costs. However the risks borne are rarely taken into account by those passing judgement and yet, almost inevitably, there will be occasions when the risk taker incurs losses. Stocks will spoil, markets will fall, cheaper imports will enter the country, consumer tastes will change, and so on. These losses can only be observed if adequate surpluses were generated in previous periods. Risk bearing must be acknowledged as a cost since what is uncertain is not whether they will occur, but when they will occur.

Market intelligence: As for as is possible marketing decisions should be based on sound information. The process of collecting, interpreting, and disseminating information relevant to marketing decisions is known as market intelligence. The role of market intelligence is to reduce the level of risk in decision making. Through market intelligence the seller finds out what the customer needs and wants. The alternative is to find out through sales, or the lack of them. Marketing research helps establish what products are right for the market, which channels of distribution are most appropriate, how best to promote products and what prices are acceptable to the market. As with other marketing functions, intelligence gathering can be carried out by the seller or another party such as a government agency, the ministry of agriculture and food, or some other specialist organisation. What is important is that it is carried out.

Links between agriculture and the food industry

The link between agriculture and food continually evolves. In primitive societies, the farmer and consumer were either the same family or close neighbours who bartered their products and services as we see in figure 1.1, but as societies develop other linkages are added. Commodity traders, processors, manufacturers who convert produce into food items and retailers, among others, are interposed between the producer and consumer. A more recently introduced link into the chain is the scientist. Scientists as breeders, plant biologists, nutritionists and chemists have made an immeasurable contribution to the development of agricultural production and food manufacture over the past 50 years. It would appear that we have passed through the age of machines in agriculture, and the age of chemicals, on to the age of biotechnology in agriculture. Biotechnology has great potential for the developing countries since it is likely to be less capital intensive and more research and know-how intensive. Thus its benefits can flow faster into the poorer countries who do not have the capital. Therefore its impact could be faster, more widespread and more significant.

As the link between food and agriculture continues to evolve, we see the emergence of an agribusiness i.e. where agriculture and food become a continuum. Multinational companies like Cargill, Brooke Bond Liebig, and Del Monte are examples of vertically integrated organisations with links all the way through from agricultural production to retailing. There is a line of argument which says that it makes sense that those who are closest should the consumer should assess his/her needs and interpret them back to the primary producer.

As disposable incomes increase, the food industry will increase the quality and diversity of the products it produces. Food manufacturers will have particular expectations of agriculture as a supplier of their raw materials, including:

Quality: To build a profitable business, food manufacturers seek to establish a preference for their products by differentiating those products in some way which is meaningful to consumers. Then, in order to enable consumers to recognise the differentiated product, manufacturers brand that product. Manufacturers can then work on building consumer loyalty to these brands. Brand loyalty is normally only established by delivering high quality consistently. As disposable incomes rise, the market tends to develop more sophisticated needs and the quality of the raw material becomes even more critical. Where agriculture is seeking to serve a food industry, that itself is seeking to meet these more sophisticated needs and wants, it can expect to face increasing emphasis on quality. Equally well, agriculture can expect to share in the better return for innovative improvements in quality.

Cost: Next to quality will come cost. With an increased capability to search the world for raw materials, the food industry is able to find the lowest cost source for any given level of quality. For the food manufacturer, the country in which he/she manufactures, or markets, need no longer be the source of agricultural produce. Improved transportation and communications mean that the world is becoming his/her source of supply. This is a significant change in the competitive environment of agriculture which the farming community has to realise, because they have, hitherto, been largely concooned in their respective domestic markets.

Non-seasonality: Agricultural products were traditionally seasonal in their production and supply. Modern technology and husbandry practices mean that food manufacturers need not have their production schedules dictated by the seasons. Indeed, the capital intensive food industry cannot afford to incur the high costs of under utilising its capacity. This means that farmers will have to complete in terms of reducing seasonality or fitting into a pattern of social competitiveness.

Reliability: A manufacturer who has invested heavily in building up his brand will be very keen to get reliable supplies in terms of quality, timing and cost. Producers of agricultural produce will be increasingly judged on their reliability in all of these respects.

Processing: Ease of processing will become an increasingly important expectation of the food industry. Like all industries, reductions in the costs of capital equipment, wages and inventories are important objectives. For example, farmers who can deliver on the 'just-in-time' principle will contribute towards reducing a manufacturer's working capital and space requirements. Farmers who can do part of the secondary processing and/or performing functions such as the post harvest treatment of the crop or transporting will be adding another advantage. Crops that are specially bred or designed to facilitate processing (e.g. seedless fruits, featherless chickens, coffee beans without caffeine, low cholesterol meats) are another type of advantage that the food industry could expect from agriculture. In short, the competitive advantage will rest with those able to add most value and can differentiate what they are offering from that of other suppliers.

Product differentiation: In competitive brand marketing, the food industry has to innovate continuously to create new products that are different from and superior to existing ones of their own or competitors. The scope of innovation has traditionally been at the processing stage. Whilst this will continue to be an important area for innovation, manufacturers will increasingly tend to look for innovative changes in the agricultural produce itself. This may be in terms of novel tastes, improved texture, more attractive shapes, etc.

Health aspects: We have already said that in the more sophisticated food markets, healthy eating can become a priority among consumers. Therefore, farmers will have to consider the health connotations of what they choose to grow. There are two aspects of health to be taken into account. First, consumers may be interested in the food itself i.e. low fat, low/no sugar or low/no salt. It would be a mistake to think that health issues are confined to the more sophisticated food markets or to the wealthier segments of the community. Nutrition is important in all segments of the market. Even where the poor receive adequate amounts of food to fend off starvation, they are often malnourished. Thus farmers have to be concerned about the nutritional value of the produce they grow. Second, the consumer may be more, or equally, concerned about the food production methods i.e. the avoidance of chemicals like herbicides, pesticides etc. This may mean a change to the farmer's husbandry practices with implications for the costs of production. The consumer and the food industry will expect the farmer to produce without potentially dangerous chemicals, but at no extra cost to them. This will be another challenge for agriculture.

Agricultural and food marketing enterprises

The principal component of any marketing system are the institutions and enterprises of which it is comprised. Three of the principal forms of enterprise to be found in developing countries are discussed in this section. These are: private companies, marketing boards and co-operatives.

CO-OPERATIVES IN THE AGRICULTURE AND FOOD SECTORS

The co-operative enterprise has its origins in the 19th century and has become one of the most ubiquitous examples forms of business/economic enterprise. Co-operatives exist in all countries of the world and operate under diverse political systems: from communism to capitalism. The majority of these co-operatives are, through their national apex organisations, ultimately in membership of the International Co-operative Alliance (ICA), the representative world body of co-operatives of all types.

The motivation to form co-operatives has three particular aspects:

- The need for protection against exploitation by economic forces too strong for the individual to withstand alone
- The impulse for self-improvement by making the best use of often scarce resources
- The concern to secure the best possible return from whatever from of economic activity within which the individual engages whether as a producer, intermediary or consumer.

It is the belief that each of these aspirations can most advantageously be pursued and secured in concert with like-minded people that provides the stimulus to co-operative action. The underpinning principles with are those of self-help, voluntary participation, equity, democracy, and a common bond of common need and purpose. The cohesion of the group is maintained by ensuring that individual members cannot secure power or gain advantages at the expense of the others. Co-operatives reward participation in the co-operative venture rather than rewarding capital Self-interest is a primary motivator in co-operative enterprises, with economic gain being the primary objective. In these respects, co-operatives differ little from capitalistic enterprises; self-interest is simply pursued in a different way from the capitalist enterprise. Thus, the rate of interest paid on share capital is fixed and limited, and not subject to variation according to the amount of profit made. Secondly the use and distribution of surplus is restricted to one or more of the following purposes:

- Allocation to reserves, where it becomes collectively-owned capital and is thereafter non-distributable
- For use on, or donation to, common-good, community project
- Distribution to members in proportion to the trade each member has done with the co-operative. In other words, the distribution is made not in relation to capital held, but by declaring a bonus or dividend per cash unit of trade done

Why agricultural diversification is vital for our rural transformation:

Diversification of agriculture refers to the shift from the regional dominance of one crop to regional production of a number of crops to meet ever increasing demand for both cash crop and food production. It takes into account the economic returns from different value-added crops with complementary marketing opportunities.

A diversified portfolio of products ensures that farmers don't suffer complete ruin when the weather is unpredictable. It manages price risk, on the assumption that not all products will suffer low prices at the same time. Unfortunately, most farmers often do the opposite of diversification by planting products that have a high price in one year, only to see the price collapse in the next.

Diversification in agriculture is key in achieving food security, improved human nutrition and increase in rural employment. Without diversification, farmers who are dependent on exports run a number of risks. A classic example is the Caribbean banana industry, which collapsed as a result of the removal of quota protection on EU markets, necessitating diversification by the region's farmers.

Farmers in several countries, including, India, Kenya, Sri Lanka, etc., have already initiated diversification as a response to climate change. For instance, government policy in Kenya to promote crop diversification has included the removal of subsidies for some crops, encouraging land-use zoning and introducing differential land tax systems.

The rate at which Uganda's soil is losing fertility, with crops drying up, calls for urgent government intervention. For instance, a recent study by International Center for Tropical Agriculture (2011) revealed that climatic suitability of most of Uganda's crops like tea will decline significantly and tea production is predicted to disappear almost completely due to poor agronomic practices. It is predicted that some areas will even become unsuitable for tea, especially in western Uganda and farmers will need to identify alternative crops.

Many farmers lack access to such vital information. I appeal to the government to start addressing key issues that have direct impact on the citizens, such as the banana wilt and coffee wilt diseases ravaging plantations in Ankole region. The government needs to set up agricultural information centres, make information available to rural farmers and provide the supportive agro-processing, market and communications facilities. There is a need for government to equip and facilitate scientists to embark on serious research, assess soil condition, and advise farmers on the right direction. Extension coordinators need to educate farmers about the values of agricultural diversification.

The case of agricultural diversification in India

The decline in the share of agriculture in India's GDP has been rapid in the post-liberalisation era; despite the fact that the growth of agricultural income during the 1990s has been marginally better than the corresponding rate of growth in the 1980s. Growth in agriculture stagnated towards the end of the 1990s and has decelerated thereafter. In this context, the composition of income from agriculture and allied sectors of the economy has been suited.

Agricultural diversification has been achieved as a larger area of land is now utilised for the cultivation of high value crops. With trade liberalisation, the relative prices of exportable commodities have increased rapidly and those of importable commodities have comparatively decreased. In the short run, a continuous increase in the relative price of a commodity enhances its production more often by substituting it for importable commodities without any statistical effect on the cropped area. As a result, the share of exportable commodities has increased in the total value of agricultural output.

Here it would be apt to rely on a study based on alternate measures (namely the Simpson index) of diversification for understanding the state of affairs and concentration of non-food crops on several factors such as income, land distribution, irrigation intensity, institutional credit, road density, urbanisation and market penetration. Regression analysis suggests that increased road density and urbanisation encourages commercialisation of agriculture and with commercialisation, farms in a region are increasingly specialised in the production of certain crops and crops-groups as per the resource infrastructure and institutions of the region.

Considering the multi-dimensional importance of agriculture diversification, it is important to understand the drivers of agricultural diversification in the country. Changes in the relative prices of corps have influenced the crop enterprise mix immensely. Price is basically a reflection of the demand and supply situation. In a closed economy, the prices that farmers receive alternately, farm harvest price is influenced by the minimum support price (MSP) and the MSP has been influencing acreage under crops. But in several growing economies, MSPs for agricultural commodities have been increased in recent past. Apart from contributing to food price inflation, this increases the spread between prices paid to producers and subsidised prices charged to consumers, fueling the fiscal burden. Since MSPs need not always extend to all agricultural commodities and public procurement need not cover all commodities either, this creates perverse price signals and distorts resource allocation.

Triggered by food price increases, there have been interventions on the consumption side, including price controls, consumption subsidies, food aid, food for work arrangements, cash transfers and the elimination of taxes on consumption across a range of countries. Are these fiscally sustainable? Do they lead to additional distortions? Do they lead to supply-side adjustments or are they knee-jerk reactions? But in India's case, agricultural diversification has not affected much through the price rise or policy response directed on that —so the case of agricultural diversification remains positive in India.

The per capita income is hypothesised to affect the diversification as measured with the percent of non-food crops in either way. The non-food crops more specifically, fruits and vegetables are increasingly recognised as a new source of growth in agricultural income. On the other hand, an increase in the per capita income is the cause of a shift in consumers' preferences from staple food items to fruit and vegetables. Such changes in the dietary pattern are the causes of a diversification in the production portfolio. This implies a positive effect of income on the percent of gross cultivation areas (GCA) under non-food crops in the country.

The size and the quality of land has always been an important factor in agricultural production relations. The average size of operational holding is often considered an important determinant of crop diversification. These variables are supposed to have a negative effect on diversification indices. Many populous states of India such as Bihar and UP are witnessing the growth of small size of farmland, which is affecting productivity and income involved with this. After an extent, the role of technology in improving productivity while alternately reducing the unit cost of production and conserving natural resources cannot be overemphasised. Thus, land management is the need of hour to sustain the welcome pattern of agricultural diversification.

The growth in agriculture during the 1990s has widened rural inequity. Leakage in the rural economy has increased with the inequitable growth of income in the system. Some of the increased leakage in the rural economy is also associated with the kind of growth of manufacturing and how the process of rural development is being handled by the (state) governments and policy agencies. Consequently, it's evident that very few policy measures are properly responding to the haunting problems in rural areas under the inefficient local administrative machinery that instead forward the pace of implementation of central plans and create many bottlenecks for stretching the project for all wrong ends.

A forward looking, speedy metamorphosis of India's bottom tier administration would be the pre-imminent step for retaining agriculture as an occupational option and further for thinking of carrying on with its bright materialistic side that comes through diversification and knowledge based farming.

The Declaration on Agricultural Diversification was launched and signed in Paris on 7 December 2015 by world renowned research leaders during the recent meeting of the United Nations Framework Convention on Climate Change (UNFCCC) COP21. The Declaration, which aims to build agricultural resilience and food security for future climates, will underpin the Global Action Plan for Agricultural Diversification (GAPAD) to be developed during 2016.

The Declaration was signed in the presence of Former Malaysian Prime Minister, Tun Abdullah Ahmad Badawi. Founding signatories included Dr Trevor Nicholls of the Centre for Agriculture and Bioscience (CABI), Dr David Molden of International Centre for Integrated Mountain Development (ICIMOD), Dr Jose Joachim Campos of Centro Agronomico de Investigacion Ensenanza (CATIE), Prof. Meine Van Noorwijk of the World Agroforestry Centre (ICRAF) and representatives from the Association of International Research and Development Centers for Agriculture (AIRCA), Food and Agriculture Organisation of the United Nations (FAO) and Global Alliance for Climate-Smart Agriculture (GACSA).

Advantages And Disadvantages Of Diversification In Agriculture Environmental Science:

Agriculture is the pre-dominant economic line of work of the rural communities in India, and plays a vital role in the socio-economic development of these communities. India acquires its major share of production revenues from the rural / agricultural sector of the economy. The agriculture sector in India is enormously significant in spite of its declining share in GDP. Sectoral shifts occurred as a result of the industrialization which had raised the Services sector shares in GDP during the 1990's, where as Agriculture, which had a major share in GDP in the 1950's, contributed only 22.5 % by the end of March 2004 (Economic Survey, GOl, 2003-04). Thus came up a pressing need for a paradigm shift in the government's agricultural policy to address the problems faced in the agricultural sector in the new domestic and global economic environment and avenues to enhance the income of the farmers. The possible solution for meliorating the agro sector is Diversification. This study traces the definition of diversification, area expansion problems, immediate needs, and its future prospects.

Diversification can also involve "a shift of resources from one crop (or livestock) to a larger mix of crops and livestock, keeping in view the varying nature of risks and expected returns from each crop/livestock activity, and adjusting in such a way that it leads to optimum portfolio of income" (Joshi, et al, 2003). It is a way of a gradual movement from subsistence staple food crops towards diversified market-oriented crops which have a larger potential for land returns. DOA being a strategy would open up opportunities, to a large extent, for value addition in agriculture and will also lead to better crop planning and improve the earning opportunities in the farm community. In India, Andhra Pradesh has been proactive in taking up agricultural diversification as a strategy to accelerate the growth of agriculture.

Crop Diversification

Crop diversification takes into account the economic returns from different value-added crops. It also implies the effective use of environmental as well as human resources to grow a mix of crops with complementary marketing opportunities, and it entails shifting of resources from low value crops to high value crops. Due to globalization, crop diversification in agriculture is also a means to increase the total crop productivity in terms of quality, monetary and quantity value under specific, diverse agro-climatic situations all over the world.

There are two approaches to crop diversification in agriculture.

Horizontal diversification - the primary approach to crop diversification used in production agriculture. In this approach, diversification normally takes place through crop intensification which means adding new high-value crops to existing cropping systems as a way of improving the overall productivity of a particular farm or a region's farming economy as a whole. Vertical diversification approach in which value is added to the products by farmers through various methods such as processing, regional branding, packaging, merchandising, or other efforts to enhance the product.

Opportunities for crop diversification normally vary depending upon the risk, opportunity and the feasibility of proposed changes within a socio-economic and agro-economic context. Crop diversification may occur as a result of government policies. The "Technology Mission on Oilseeds", "Spices Development Board", "and Coconut Development Board" etc is some examples where the Indian government created policies to thrust changes upon farmers and the food supply chain at large as a way of promoting crop diversity.

Crop diversification is the outcome of several interactive effects of many factors

Environmental factors which includes irrigation, rainfall, and temperature and soil fertility.

Price-related factors which includes output and input prices with respect to national and international trade policies and other economic policies that affect the prices either directly or indirectly.

Technology-related factors which includes seeds, fertilizers and water technologies, but also those related to marketing, harvest, storage, agro-processing, distribution, logistics, etc.

Household-related factors which includes regional food traditions, fodder and fuel as well as the labor and investment capacity of farm people and their communities.

Institutional and Infrastructure-related factors which includes farm size, location and tenancy arrangements, research, in-field technical support, marketing systems and government regulating policies, etc.

All these five factors are interrelated.

Area expansion problems under rice and wheat crops

Scaling up production area poses several new problems of significance such as:

1. Excessive use of groundwater leading to poor water use efficiency and depletion of groundwater.
2. Deterioration of soil health or soil fertility.
3. Multiple infestations of weed flora, insect pests and diseases.
4. Indiscriminate use of energy such as chemical, electricity or disease, etc.
5. Reduction in the availability of other protective food and high value crops.
6. Pollution of agro-ecosystems.

Despite the expansion problems, crop diversification has the potential to be an economic driver in agricultural regions. It may prove to become the paramount importance in meeting challenges that arise from a post-green revolution scenario. In view of shrinkage of agricultural land and operational holdings due to expansion of urban centers, changes in consumer food habits, exponential population growth rate, farmers are pressured to include or substitute additional crops in to the cropping system.

Reforms in agriculture marketing in India

India is an agricultural country and one third of its population depends on the agriculture sector directly or indirectly. Agriculture remains as the main stray of the Indian economy since times immemorial. Agricultural Marketing is mainly the buying and selling of agricultural products. In earlier days when the rural economy was more or less self-sufficient, the marketing of agricultural products presented no difficulty as the farmer sold his produce to the consumer in cash.

Importance and Objectives of Agriculture Marketing:

The farmer has realized the importance of adopting new techniques of production and is making efforts towards enhancing his income and standards of living. As a consequence, the cropping pattern is no longer dictated by what he needs for his own personal consumption but what is responsive to the market in terms of prices received by him. Agricultural marketing plays an important role not only in stimulating production and consumption but also in accelerating the pace of economic development.

The agriculture marketing system plays a dual role in economic development in countries whose resources are primarily agricultural. Increasing demands for money with which to purchase other goods leads to increasing sensitivity to relative prices on the part of the producers, and specialization in the cultivation of those crops on which the returns are the greatest, subject to socio-cultural, ecological and economic constraints. It is the marketing system that transmits the crucial price signals.

The Government of India has been playing an important role in developing agriculture marketing system in the country. The marketing division of the Department of Agriculture, Cooperation & Farmers' Welfare is entrusted with the implementation of policy and programme related to agricultural marketing. The agriculture sector needs competitive and well-functioning market for farmers to sell their produce. In order to remove respective and monopolistic practices of present marketing system, to reduce the intermediaries in supply chain, to reduce wastage by way of promoting integrated supply and value chain and to benefit farmers through access to global markets, reforms in agricultural markets have to be an ongoing process.

Integrated Scheme for Agricultural Marketing (ISAM):

In the Ministry of Agriculture and Farmers Welfare, Marketing Division is also implementing ongoing Central Sector Schemes from XII plan, which have been integrated into a new scheme viz. the Integrated Scheme for Agricultural Marketing. ISAM has six sub-schemes namely:

Agricultural Marketing Infrastructure (AMI)

Marketing Research and Information Network (MRIN)

Strengthening of Agmark Grading Facility (SAGF)

Training, Research and Consultancy through Choudhary Charan Singh National Institute of Agricultural Marketing (NIAM)

Agri-business Development through Venture Capital Assistance (VCA) and Project Development Facility

National Agriculture Market

For facilitating better value of the produce by the farmers, accessible, competitive, efficient and barrier free markets with adequate marketing infrastructure and real time market information is being considered instrumental to supplement the Government's effort towards doubling farmers' income in a time bound manner.

As a part of the reform agenda, with the objectives to create barrier free market, enhance competition & transparency in transactions and widen choice to farmers for sale of their produce, Prime Minister Shri Narendra Modi launched the pilot of e-NAM - the e-trading platform for the National Agriculture Market (eNAM) on 14th April, 2016. Initially 21 mandis in 8 states had been linked to the National Agriculture Market.

Presently, 585 mandis are linked with the e-NAM platform. For this purpose, each mandi will be provided assistance upto Rs 75 lakhs for their basic infrastructures (sanity grading and packaging). In addition to this, it has also been suggested to denotify from APMC Act (Mandi Act) those commodities which are quickly perishable.

Keeping in view the need of making marketing of commodities easier for farmers, the Government of India has launched six new user friendly features of National Agriculture Market (e-NAM) platform. i.e. (i) e-NAM Mobile App; (ii) BHIM payment facility; (iii) New and improved Website with eLearning Module; (iv) MIS Dashboard; (v) Grievance Redressal Management System for Mandi Secretaries; and (vi) Integration with Farmer Database.

In the National Agriculture Market scheme, only those States / UTs are eligible to link their markets to e-NAM portal, which have undertaken reforms in their APMC Acts in respect of (i) e-trading; (ii) single point levy market fee across the state; (iii) single unified trading licence valid across the state.

Government has taken several steps to provide remunerative prices to farmers for their produce which include procurement through designated procurement agencies including price support scheme (PSS) for pulses, oilseeds and cotton, implementing e-National Agriculture Market (e-NAM) and enacting a model Agricultural Produce and Livestock Marketing (Promotion & Facilitation) Act, 2017. A Committee under the Chairmanship of Additional Secretary (Marketing) was constituted to draft it, which submitted the draft of new Model Act in December, 2016. Model Act submitted by the Committee was sent to the States and Union Territories for comments. Besides, it was also uploaded on the Agriculture Department's official website for comments of general public. The final form of model APLM Act, 2017 has been prepared after incorporating the comments of all stakeholders which was inaugurated on 24th April 2017 for the follow up of the states. The progress of reforms with respect to 7 vital areas identified by the Department is:

Setting up of markets in private sector;

Direct marketing (direct purchase of produce from farmers by processors/ exporters/bulk buyers, etc outside the market yard);

Farmer - Consumer markets (direct sale by farmers to consumers) to be set up by a person other than a Market Committee;

Contract Farming;

E-Trading;

Single point levy of market fee across the State;

Single trading license across the State;

De-linking of Provisions of Compulsory Requirements of Shop/Spaces for Registration of Traders/Market Functionaries;

Take out Fruits and Vegetables out of APMC Act.

Recently, Government of India has announced minimum support prices (MSPs) for 22 mandated crops including paddy and wheat and fair and remunerative price (FRP) for sugarcane on the basis of recommendations of the Commission for Agricultural Costs and Prices (CACP).

7

FINANCIAL MANAGEMENT FOR AGRIBUSINESS

Financial Management for Agribusiness presents a practical approach to financial decision making for all those involved in agribusiness, including farmers, horticulturists and supporting businesses, to manage invested funds, physical resources and labour. It covers all the stages leading to a completed business plan and provides straightforward worked examples for each step.

Most farmers and ranchers still prefer to concentrate on production rather than planning, recordkeeping and marketing. However, the escalation in input costs, the ability to access capital, and greater volatility in commodity markets make financial and risk management skills very important. Reductions in the government farm program safety net, advances in biotechnology and information technology, and the increasing globalization of competition will continue to strain traditional approaches to management as we move further into the 21st Century.

Many farmers spend thousands of dollars to improve production, but hesitate to spend money on tools that can help them improve their financial management. In fact, managing farm finances offers the best possibility for adding profits and ongoing profitability to an agricultural operation. Consider these reasons why financial management is crucial to the success of your agriculture operation.

Those who succeed will manage their operations as farm/ranch businesses, with more emphasis on recordkeeping, planning, profiability analysis, and repayment-based financing, and with better monitoring of production and business processes. While production can not be neglected, it should be balanced with attention to other key performance areas.

The purchase of inputs should be based on economic analysis, not just increased output. Successful producers will need more information about financial analysis, marketing, analysis of alternative enterprises, and the "bottom line" contribution of each existing enterprise.

This shift applies equally to lending institutions. Lenders will have to undertake more in-depth financial and risk analysis. "Equity/collateral" and "character" lending protect the lender, but if the producer can not generate sufficient profits and cash flow to service debt, the lender is doing a disservice to the producer by providing capital that may end up actually increasing risk and jeopardizing the producer's operation and equity.

As a result, agricultural producers and their lenders must adopt some of the basic analytical and planning tools used in business. To start with, a producer should ask and honestly answer the following questions:

Is this an efficient production system?

Can commodities be produced below the average cost of production?

Is there a marketing plan based on adequate market analysis and the break-even price for each commodity being produced?

Is there an accurately prepared business

plan based on economic considerations?

Does the business have sufficient equity and liquidity to handle adversity or a cyclical downturn?

Is the operation large enough to provide the required family living withdrawal?

If not, is there sufficient off-farm income?

Have the risks involved in the business been adequately analyzed and compensated for?

Is there an adequate system for obtaining management information and monitoring business performance?

Questions like these should be addressed continually, and not just during an initial planning process.

DEVELOPING FINANCIAL MANAGEMENT SKILLS

Answering these questions will require a thorough analysis of the operation and may require financial and marketing management skills the producer does not yet possess or practice. Many producers will need to embark upon an intensive educational program to gain the essential farm/ranch business management skills and the confidence to use them successfully. Producers who want to succeed need to develop a farm/ranch business plan that outlines, in detail, how they plan to generate a profit and manage business risks. They should start by outlining specific business/personal goals and objectives and preparing complete inventories of resources available. That is, they can not prepare a meaningful plan unless they specifically outline what they wish to accomplish and the basic "resources" available to accomplish this "objective."

The resource inventory should cover all areas of the business and include such items as farm/ ranch maps by productive capacity, soil analysis, water resources, machinery, labor skills available, management capabilities, detailed crop/ livestock production history, analysis of market information, and complete financial statements.

The Planning Process

With the data obtained in the first step of the planning process, a producer can begin developing and comparing alternative production, marketing and financing plans. The goals indicate business and personal priorities and help establish performance objectives. Data on the physical resources (including the land, labor, tangible working assets, and capital), the institutional considerations, and the managerial capacity (strengths and weaknesses) suggest the range of enterprises, production systems, and marketing tools to consider in preparing a comprehensive whole farm/ranch plan.

The next task is to develop a plan the operator will follow for several years. A long-term plan integrates the production, management, marketing and financing activities that will be required to accomplish the operator's goals. This plan may require a significant change in the current business structure. At a minimum, the plan should include:

A complete resource inventory and a

specific outline of goals and objectives.

A 5-year crop/livestock rotation plan. Enough years should be included in the plan to move through any transition in production and financing that may cause significant fluctuations in the year-to-year cash flows.

Capital improvements needed immediately (first planning year), in the intermediate term (1 to 5 years), and long term (5 to 10 years).

Capital needed for operating the business and for the long term.

A physical production plan by commodity and for the whole farm/ranch. A calendar of events by month and by year, outlining all of the required production practices and resources, should be prepared for each commodity and for the whole farm/ ranch.

A long-term strategic assessment based on both historical data and current outlook information. This should include an assessment of government farm programs, the general economic outlook (e.g., inflation, interest rates, taxes), specific industry/commodity outlook (trends and cycles), and technological developments.

Individual enterprise budgets and a consolidated whole farm/ranch cash flow projection by month for at least the first year, and by year for the following years.

A marketing plan for the contemplated enterprises. The marketing plan should include sensitivity analysis for at least three price levels: expected, pessimistic and optimistic.

Although the items mentioned above are listed separately, they are actually part of an integrated planning process. For example, the resource inventory should help to identify potential farm/ranch enterprises. A budget forecast for current and potential enterprises, using the three alternative price scenarios, should identify alternatives with the greatest potential profitability and those whose profitability is most uncertain. Scarce resources such as capital, land and labor should be allocated to the most profitable enterprises. A 5-year crop/ livestock rotation plan would increase the probability of sustaining maximum economic yields. By combining operating and long-term capital requirements into a multi-year cash flow projection, a producer can better evaluate the financial feasibility of short-, intermediate- and long-term plans. Balance sheets and projected income statements will be needed to evaluate net worth, profitability, economic efficiency and financial structure.

If not being done already, detailed enterprise and whole farm/ranch records should be kept from the time the plan is implemented. Financial statements and cash flow projections with "what if" analysis should be prepared and adjusted as often as necessary to analyze risks and keep the plan on track. Successful business management requires continuous monitoring and control.

Financial Statements

A farm/ranch recordkeeping system facilitates the preparation of financial statements and the monitoring of business performance. Properly prepared balance sheets, income statements and cash flow projections will be needed to help the manager keep the business on track. This is done by comparing actual performance to the plan and taking action when necessary. Farm businesses have failed simply because the farmer or his lender recognized problems and took corrective action too late.

Financial Analysis of a Farm or Ranch Operation

High among any business's goals should be profitability, liquidity, solvency and efficiency. Yet, some managers who are very capable in the technical and production aspects of their businesses are completely in the dark when it comes to measures of business status and performance. They leave their recordkeeping to bookkeepers, or worse, use the "shoebox" record system. The only "business analysis" they do is to sign their tax forms or prepare "required" financial information for their lenders.

Effective managers are able to determine the financial position and performance of the business at any time. They can evaluate where the business is, has been, and is going, which helps them control the business operation over time. They can develop sound operating plans and arrange for credit needs.

The basic tools needed to conduct farm/ ranch business analysis are adequate records and properly prepared, complete financial statements.

Farm and Ranch Financial Statements

The most useful coordinated financial statements to use for summarizing a farm's or ranch's financial position, and for demonstrating managerial ability, are the balance sheet, income statement and cash flow statement. These financial statements allow a producer to systematically analyze financial progress, plan operations for the year ahead, and demonstrate creditworthiness to lenders. To be most useful, these statements should be compared over several years.

If not available from previous years, there is no better time to start preparing these statements than now.

The Balance Sheet

The balance sheet, also referred to as the statement of financial position or net worth statement, is a summary of all assets and liabilities of the farming or ranching operation at a specific point in time. At a minimum, a balance sheet should be prepared as of the last day of the fiscal year.

The Income Statement

The income statement reflects the profitability of a farm/ranch business over a period of time, usually from the beginning to the end of the tax year. For accurate analysis, it is critical that income be calculated on an accrual-adjusted basis. This does not imply, however, that farmers/ ranchers should not use the more flexible cash basis for tax accounting. It does mean that cash basis income accounting for management purposes can give a very distorted and inaccurate picture of business performance, and can produce lags of up to 2 years in recognizing profitability problems that may be developing.

The Cash Flow Statement

A projected cash flow statement is a listing of all anticipated cash inflows and outflows for a specified future period. It includes both farm/ ranch and non-farm income, and all projected cash outflows, including operating expenses and capital outlays, family living expenses, borrowing transactions and tax payments.

Using and Analyzing Financial Statements

Farmers and ranchers should use the information provided by the balance sheet, income statement and cash flow statement to measure and understand business performance. Only by analyzing the components of all three financial statements and their interrelationships can a clear picture of financial position and performance be developed.

There are three warnings that reinforce the importance of using all three statements.

The first is that a business can be going broke and still be generating a positive cash flow for a period of time by deferring expenses, restructuring or refinancing debts, selling off assets (including inventories), and not replacing capital assets as they wear out (i.e., living off of depreciation).

Second, unless 1) the operator is generating enough off-farm income, 2) assets are appreciating faster than the business is losing money, and/or 3) the operator is inheriting sufficient money (assets) to offset losses, a business has to be profitable to survive in the long run.

Third, many farmers and ranchers have failed even though the business was profitable because they were consuming more than the business was earning.

After a careful review of all factors involved, the farmer/rancher may decide that something needs to be done. Unfortunately, there is no single prescription that will work for every operation. Like a medical problem, the solution will depend on a specific diagnosis based on the business and the individuals involved. Financial statements that are carefully prepared and analyzed will help the manager see warning signs in time to make needed adjustments. Good financial analysis does not guarantee success, but it definitely improves the odds.

Financial Management

Lack of financial management is a primary cause of failure for value-added businesses.

Financial management starts with keeping records. There is an old saying that "you can't manage what you can't measure." Records provide you the data needed to measure the activities of your business. In other words, records are the foundation for the financial management of your business.

After you have created a good set of records, you can begin constructing financial statements. These statements focus on the three main aspects of financial management: profitability (are you making money), liquidity (do you have enough cash) and solvency (how close are you to going broke). Financial statements tell you your business's current financial status along with its past financial status. Historic statements can identify financial trends that will help you project the future financial status of your business.

A good set of records can also help you budget future potential profitability of various changes in your business. These budgets help you make meaningful decisions about the future course of your business.

If your business is experiencing financial stress, financial analysis and management can help you identify and analyze alternative scenarios for relieving financial stress.

For more information on this topic, see the links listed below of articles posted on related Web sites. Also see the worksheets and calculators available on the Business Workbench.

- An Integrated Financial Management Package - Kansas State University Extension.
- Financial Management; The Key to Farm-Firm Business Management - Texas A & M University Extension – Farmers must adopt some of the basic planning and management tools used in business.
- Is Your Company Nudging the Numbers? - HBS Working Knowledge – Pressure to meet the numbers is greater than ever. But don't let your company issue a misleading earnings report.
- Analyzing Your Current Financial Position - The U.S. Chamber of Commerce – Delves into some of the more sophisticated ways of examining your financial statements and other aspects of your business, to identify trends, spot problems before they become too large and compare your business to others in the same industry.

- Preparing Financial Projections - PlanWare – Financial modeling with a computer, how models work, planning to plan, and tips and traps.
- Important Farm Business Terms Defined - Finance – Kansas State University Extension – Provides definitions of important farm business terms.
- Financial Management in Agriculture

The use of sound financial management practices in agriculture is quickly becoming necessary for survival and are a required for outstanding economic performance. We developed a research program to determine how widely farmers had adopted various financial management practices and whether these practices had any impact on farm profitability. Not surprisingly, we found that many simple financial management practices had a strong relationship to farm financial performance. Among other things we found that farmers who took the time to prepare a written or computerized cash flow analysis of their investment projects were much more profitable than their peers who either did not conduct the analysis or those that "did it the analysis in their head". Check out our financial management scorecard to see how you or your borrowers stack up.

Importance of agriculture finance

Finance in agriculture is as important as development of technologies. Technical inputs can be purchased and used by farmer only if he has money (funds). But his own money is always inadequate and he needs outside finance or credit. Professional money lenders were the only source of credit to agriculture till 1935. They use to charge unduly high rates of interest and follow serious practices while giving loans and recovering them. As a result, farmers were heavily burdened with debts and many of them perpetuated debts. There were widespread discontents among farmers against these practices and there were instances of riots also. With the passing of Reserve Bank of India Act 1934, District Central Co-op. Banks Act and Land Development Banks Act, agricultural credit received impetons and there were improvements in agricultural credit. A powerful alternative agency came into being. Large-scale credit be cameavailable with reasonable rates of interest at easy terms, both in terms of granting loans and recovery of them. Although the co-operative banks started financing agriculture with their establishments in 1930's real impetons was received only after Independence when suitable legislation were passed and policies were formulated. There after, bank credit to agriculture made phenomenal progress by opening branches in rural areas and attracting deposits.Till 14 major commercial banks were nationalized in 1969, co-operative banks were the main institutional agencies providing finance to agriculture. After nationalization, it was made mandatory for these banks to provide finance to agriculture as a priority sector. These banks undertook special programs of branch expansion and created a network of banking services throughout the country and started financing agriculture on large scale. Thus agriculture credit acquired multi-agency dimension. Development and adoption of new technologies and availability of finance go hand in hand. In bringing "Green Revolution", "White Revolution" and now "Yellow Revolution" finance has played a crucial role. Now the agriculture credit, through multi-agency approach has come to stay. The procedures and amount of loans for various purposes have been standardized. Among the various purposes "Crop loans" (Short-term loan) has the major share. In addition, farmers get loans for purchase of electric motor with pump, tractor and other machinery, digging wells or boring wells, installation of pipe lines, drip irrigation, planting fruit orchards, purchase of dairy animals and feeds/fodder for them, poultry, sheep/goat keeping and for many other allied enterprises.

AGRICULTURAL & RURAL FINANCE

Rural finance comprises the full range of financial services - loans, savings, insurance, and payment and money transfer services - needed, offered, or used in rural areas by household and enterprises. The term encompasses agricultural finance.

Agricultural finance refers to financial services ranging from short-, medium- and long-term loans, to leasing, to crop and livestock insurance, covering the entire agricultural value chain - input supply, production and distribution, wholesaling, processing and marketing.

Rural and agricultural financial services are provided by formal and informal financial institutions as well as through financial arrangements within the agricultural value chain.

While the majority of Africa's population lives in rural areas and depends on agricultural production, the supply of financial services to the sector is inadequate, with, on average, a mere 5 percent of domestic resources being allocated to the agricultural sector.

Reasons for the lack of access to finance in rural areas and in the agricultural value chains are numerous. They can be found in the slow and uneven entry of formal financial institutions into rural areas, which leads to rural clients often remaining beyond the reach of financial outlets, to the reluctance of financial institutions to provide financial services to agricultural and rural activities, whose risk profile is frequently not fully understood and which are often informal in nature.

Factors such as poor infrastructure and widely dispersed populations in rural areas raise transaction and information costs, thus further hindering the spread of financial services. In addition, title and property rights can be difficult to verify in rural areas, posing problems in the use of collateral. Subsidized lending programs for rural recipients have also contributed to obstructing the development of a sustainable rural banking sector in Africa.

Farmers and agricultural companies typically face seasonal income and long maturation periods and are exposed to considerable risks. Seasonality requires specifically tailored financial services and conditions, such as longer repayment and grace periods, less frequent repayments, or leasing products. Agricultural risks to be considered include price fluctuations for inputs and products or crop failure due to pests and diseases, temperature or variable rainfall.

Despite these difficulties, formal rural and agricultural finance has been making advances in the continent, with innovative financial services and improved risk management on both the client and institution sides. The most promising approaches include flexible credit schemes, value-chain finance, insurance products, promotion of financial literacy and the use of new technologies.

Prominent examples are the introduction of index-based weather insurance schemes in Malawi, financial literacy campaigns in Ghana, or the spread of mobile banking in Kenya. Furthermore, rural and agricultural clients continue to rely on financial institutions, like savings and credit cooperatives or village banks.

8

CHALLENGES AND OPPORTUNITIES FOR AGRIBUSINESS DEVELOPMENT

Indian agribusiness is being seen as an important arena for creation of employment and income for large sections of India's rural and agrarian population in the coming years and decades by public agencies, nationally and globally. This is happening in the context of strong agribusiness sectors in the neighbouring countries of China and those in the South-East and East Asia. This article examines the global context of agribusiness competitiveness from an Indian perspective, including the global trade regime and focusses on a few key sectors like dairying, cereals, fruits and vegetables and identifies major factors in their export performance from a value chain perspective. It also examines key global trade policies at the World Trade Organization (WTO) level and Indian agribusiness policy domain to explore ways forward to focus on promising sectors with an appropriate policy and practice steps.

India Agribusiness Market research identifies that the competition continues to intensify year-on-year with emerging applications and widening product portfolio. This report covers the 2019 scenario and growth prospects of the India agribusiness market for 2016-2026. To calculate the market size, revenue from the market sales of agribusiness products to retailers, wholesalers, and institutional buyers is considered.

The global agribusiness market is poised to register a growth of 2.8% CAGR over the forecast period. The ongoing trend towards modernization of agribusinesses such as the Trans-Pacific partnership, Agriculture Innovation program and other practices are supporting countries worldwide to strengthen their agribusiness markets.

India Agribusiness companies, strategic and financial investors are focusing more on the evolving value chain and are investing significantly across segments of the food and agriculture value chain. Agricultural input manufacturers, retailers, farmers, food processors, food manufacturers, and food retailers continue to emerge rapidly across the value chain amid strong growth prospects of the market. Further, India companies are focusing on synergistic acquisitions to boost their market shares and profitability of agribusiness.

India government is encouraging investments in agribusiness both to promote effective strategies of improved food security and as a vital source of economic development. Emphasis is given for good agricultural practices, prescriptive agronomic recommendations, data-based farming, and other precision farming applications.

To ensure that the delivery of dairy products is in line with customer expectations, major competitors in the dairy market are adopting customer engagement strategies, flexible pricing, value-added products and strengthening their distribution channels.

Amidst growing profitability in agribusiness, India companies are focusing on superior farm management and realigning their goals and available resources to market their agriculture products at affordable prices to targeted customers.

The India Agribusiness Market, Size, Share, Outlook and Growth Opportunities 2020-2026 presents a comprehensive analysis of the country's agribusiness. Key trends and critical insights into India agribusiness markets along with key drivers, restraints, and growth opportunities are presented in the report.

India grains, India sugar, India Dairy, India livestock markets are analyzed and forecast to 2025. Further, India agriculture production including wheat, corn, Barley, Rice, Sugar, Milk, Butter, Cheese, whole milk, poultry, beef, and pork production is forecast from 2016 to 2026. Demand for all these agriculture commodities is also forecast during the period.

India agribusiness market is compared against five of its competitive markets in the region to analyze the role of India on the regional front and benchmark its operations.

Global agribusiness, Asia Pacific, Europe, Middle East Africa, North America, and Latin America agribusiness market outlook is also presented in the report to provide a global perspective of the industry.

India population and economic outlook are also presented in the report to provide insights and forecasts of macroeconomic factors shaping the future of

India agribusiness markets. Further, business and SWOT profiles of three of the leading agribusiness companies in India re detailed in the report along with recent developments and their impact on overall market growth.

With much more sensors being invented and used in the world, we began to listen, measure and know things which we ignored. More traceability, more sensory analysis, there will go in agribusiness with one of the challenges from the 7, of significant importance and predominance: THE AVATAR's KINGDOM. A generation of customers and consumers become more sensitive and sensitized by the information for the sustainability on all levels. The Coca Cola Company has examined the lots of orange juice, that takes to process your refrigerants, and residues of Carbendazin are detected, for example. This defensive is used to control black spot in orange.

The orange becomes not esthetic, but the healthful aspect goes to count. And if there is anything that Coca Cola, and all processors and retail networks of the world, do not want is to have trouble with complaints and health risks of their customers. Therefore, the discussion is established, the technicians will discuss, if 30 parts per billion in orange, is harmful or not, and why that Japan accept 3,000, and Europe 200 parts per billion, and what to do with The United States that rule is zero or 10, which is practically zero? An increased challenge is exactly that of the sensors and the expansion of sensibility and very low thresholds to the question of human and the planet health.

The second challenge is about the agricultural lands. The world must double food supplies and renewable energy. FAO's data show that 70% of this increase will come from the technology, 10% of the intensity of agricultural cycles in the same area and the remaining 20%, Yeah these – from new agricultural areas. It means, the land became too worth gold. The expansion of new frontiers has been limited and the relation of ownership of agricultural lands has different story, one of the most valuated active in the world over the last 5 years. Brazil is beside the Sudan in Africa, are the two biggest reserves on the planet, in this regard.

The third challenge is the population: people. The population grows in proportion to of 4 new births per second. We will be 9 billion, we already are 7 billion. And this new habitant is fully connected and interconnected. The Internet expands the situational ethics of the past, and in addition to cellular phones connection, computers, televisions, it is coming now the new Internet of things. It mean, in the shower, microwave, on the stove, on the car, etc. The fourth challenge is not only food production, which should double, but their quality. The raw materials produced from the field begin to be more import perceived in the analysis of the marks of processors and distributors of food, fiber, energy, protein and other derivatives of agribusiness. Food and energy, their production, distribution and perception of the interconnected markets become a new challenge for the management of the entire value chain, from the first, passing through the inside and by post farms gate.

The fifth challenge, the innovation, the adoption and the management of technology and its life cycle. The problem is not count with an advanced technology, is that we need to have over than one. This diversity technology advances will require of the farms, industry, chemistry, mechanics, genetics, and processors a level of "on-line" dialog. The academic area, school and public research cannot resist to the absence of speed as well as agribusiness governance will be more based on the sectorial advice. In the technology, the stock of local germoplasms will be a factor of risk management and the environmental security, political and national interest.

The 6th challenge is the question: Who will be the farmer in 2022? In the next 10 years, who will be, as will be, how many will exist? Which segments, the clusters? Which the human competences of this new situation of producers for the next 10 years? The present became the result of the future. What we have now, which signals that we captured in this moment that show the journey and the window for tomorrow. Anticipating the movements is a key challenge for all the agents involved in agribusiness, and that will be fundamental to the construction of "brands" that resist the time. Product is what we do in the industry or on the field; brand is that we built in the mind passionate human. And if the producer is the universe to be reviewed, including the unpredictability of the world cyclical, where a European crisis, for example, we begin to see an contrary exodus: urban abandon the cities and start new businesses in the field; the old "farms" are now seen as sources of quality of life and also, entrepreneurship; If all that is a set of questions that most intrigue the current executives of agribusiness managers, here comes the 7th challenge: Marketing. The good old marketing: agribusiness without marketing is just agriculture, no deal, which means to place the human mind of all the stakeholders at the center of the table of business decisions and of the agribusiness. How much we know of this army of new customers arising from the base of the pyramid planetary? Of these current 7 billion, almost half are new entrants. If the pasta, for example, has capillarity and penetration in a 100% of Brazil's households, and we are the 3rd largest world market on the masses, is now, with the ascension of C class, almost 100 millions of consumers greedy for the consumerist wishes that began to have the penetration of new stuffed pasta. The "spaghetti" to the cannelloni is the journey that is already in fast penetration. And, as in meat, milk, vegetables and fruits, fiber, sugar, cocoa, rice ... or rubber trees.

Land, people, food and energy, the Kingdom of Avatar, technology, the producer of the future, and the marketing. The synthesis of the 7 challenges of the agribusiness.

The dialogue with society is a factor "sine qua non" the part of all agents involved on agribusiness, and build company with a new pedagogical intelligence becomes vital. In the old education we were first driven to think, and then do and if get one day in the life: feeling. The playful expression, modern, begins with the art of to do, in doing feel, feeling we think. This prudence is sacred for not fall in large errors of the decisions of leaders well-intentioned, but who fall in the sin of distractions.

The speed changed, is nearly instantaneous, and all of these seven macro challenges of the agribusiness must be packaged for the high speed of a new hybrid: do, feel and think.

In the global economy, most developing countries face several structural problems in production and consumption systems (Clark, 2007; Akenji & Bengtsson, 2014; Govindan, 2018). Besides, the literature mentions that the developing countries are associated with food security issues, poverty, unemployment, and quality of education (Larson et al., 2007; Etana & Tolossa, 2017; Rachmawati et al., 2018). However, since the nation confronted a trade-off due to scarcity, it forces the government to prioritize the sector to be addressed. Some evidence suggests that increasingly prioritize food, and agricultural affairs will lead to higher welfare, even by implementing protection strategies that tend to be overrated (da Silva et al., 2018; Inwood, 2017).

The success and failure of the government policy in agriculture sectors can be seen by the enhancement of farmer's wellbeing (TerAvest et al., 2019; Sedyati et al., 2019). The development of agriculture is not merely oriented towards increasing the physical production of various agricultural commodities, plantations, livestock, and fisheries.

The criteria for success should be measured by improving the level of income of farm households, increasing labor productivity, and improving macro- indicators such as reducing poverty and unemployment (Olowu et al., 2019). In Indonesia's context, the government has set a long-term goal to accomplish community welfare (Thufail, 2016). It consists of several efforts for improving the ability of farmers and their supporting institutions, securing food security, increasing productivity, production and competitiveness of agricultural products, and supporting forests for business diversification and support food production (Amrullah et al., 2019).

The availability of food in sufficient quantities, adequate quality, and affordable prices by the population are some of the targets to be achieved in the formulation of national food policies. Food supply instability, volatile food prices (rice) in Indonesia have been proven to trigger social instability (O'connor et al., 2017; Wildana & Alhabshi, 2018). Food security is a condition of fulfilling food for households, which is reflected in the availability of sufficient food, both in quantity and quality, safe, equitable, and affordable. The approach of the framework for the study of food security and nutrition of the community is focused on exploring aspects of food fulfillment by the community, the level of food availability, and food accessibility and stability (Saediman et al., 2019).

Healthy lifestyles and green behavior have become a new trend, and leave the old lifestyle that uses non-natural chemicals, such as fertilizers, synthetic chemical pesticides, and growth hormones in agricultural production (Rana & Paul, 2017). Such a healthy lifestyle has been institutionalized internationally, which requires a guarantee that agricultural products must have food safety attributes, high nutritional attributes, and eco-labeling attributes (Hasanah et al., 2019). These attributes are closely attached to organic agricultural products. Organic farming is a solution to restore the health of the land and the products produced (Marja et al., 2020). The people's desire to switch to organic products causes the demand for organic agricultural products to increase rapidly. Organic agriculture producers in East Java currently are still minimal. This is an opportunity for agribusinesses to develop organic food products. For this matter, this organic food phenomenon needs to be captured as an opportunity for agricultural development in East Java, especially regarding strategic food products.

The application of organic systems to food commodities, especially rice, is still a dilemma between efforts to increase production by using fertilizers or chemical pesticides as well as efforts to preserve nature that tries to control the agrochemical products (Ali et al., 2019). Organic rice agribusiness with high costs and based on some shortcomings do not cause all parties to provide support to organic agriculture. Organic agricultural products are currently only being enjoyed by certain groups, namely, the middle to the upper-class economy, because the price is more higher than conventional agricultural products (Kim et al., 2014; Altarawneh, 2016). Another obstacle to organic agriculture is related to the condition of food needs that are still importing food, so it needs better thinking when implementing organic agriculture to produce healthy, affordable, and adequate food products for East Java. For these reasons, this research is focused on the prospect of organic rice agribusiness development to support the food security program in East Java, which covers the technical and implementation of organic food systems, economic structures, and institutions, the feasibility of farming and strategies for developing organic rice commodity agribusiness.

CONTRACT GROWING

Contract growing is used primarily in the industrial production of hogs and chickens. In this arrangement, a corporation that owns livestock contracts with farmers to raise the animals to maturity. Without a contract, a farmer must manage slaughtering, processing, and distributing as well, all while keeping sale prices low enough to compete in a corporate-controlled marketplace. From afar, it might seem like a good deal.

Here's how a contract growing system works:

- A major food corporation delivers to the farm both feed and a large number of immature animals. (In the case of chickens, the animals often arrive on the same day that they are hatched.) This allows the corporation to control the breed of the animals and the content of their feed.
- The farmer raises the animals for a pre-determined amount of time, after which the company collects, slaughters, processes and distributes them.
- The corporation returns with another round of animals and feed, and the cycle repeats.

While this system guarantees farmers a market for their animals, it also burdens them with tremendous financial risk. Farmers must pay for all feed, and absorb any financial loss from animals that die. Farmers must invest considerable capital to build structures for housing the animals, a debt they carry for years, even decades. Once they own the buildings, they must pay for maintenance, utilities and insurance costs. Farmers are also responsible for disposing of the animals' waste, which means either building receptacles for storage, selling it as fertilizer, or renting land where it can be spread.

Throughout the process, the corporation maintains full control over the animals involved. If a farmer becomes "non-competitive" (meaning he or she can't raise the animals cheaply enough), the corporation may choose not to renew the contract. But because banks will not renew loans unless a farmer has a contract, getting dumped by one corporation leaves no option but to sign up with another. Too long spent with empty barns and unpaid mortgages will drive a farmer into bankruptcy.

Because corporations rarely define exactly what "staying competitive" entails, farmers often find that the only way to survive is to cut every possible cost. Too often, that includes the costs involved with responsibly caring for the animals and appropriately handling their waste. The helpless farmer is reduced from a steward of the land to someone who manages an industrial process.

Vertical Integration

One way that major corporations maximize profits is by controlling all stages of the production and distribution processes – otherwise known as vertical integration. In addition to contracting with farms to raise their livestock, corporations will own a feed company, a farm supply company, and a processing and distribution company. This allows them to profit from every level of food production without investing in permanent assets like land or losing money to unpredictable elements such as animal mortality.

Vertical integration also gives corporations enormous control over meat and milk prices, in a fashion similar to trusts and monopolies. They buy from farmers at rock-bottom prices, then use the cheap product to drive prices down and force out competitors, which consolidates their power even more.

In theory there is nothing wrong with vertical integration, as long as corporations are operating within the boundaries of the law. After all, any business should have the right to adopt methods that maximize profits. But in agribusiness, the system has undeniable consequences. It forces farmers into debt, supports farming practices that harm the environment and abuse animals, and compromises the quality of the food we eat.

9

THE IMPORTANCE OF PLANNING IN AGRIBUSINESS

Future outcomes are a function of today's decisions. Although there is a high degree of randomness and uncertainty associated with the future, you can increase the probability of a successful outcome by planning ahead. This is true in nearly every aspect of our lives, both personal and professional. For those who operate their own businesses, planning becomes increasingly important because the personal and professional aspects become more difficult to untangle. In agricultural businesses, planning may be even more vital because of the inherent uncertainty associated with agricultural production. Some important sources of uncertainty include production risk, price risk, financial (or interest rate) risk, and changes in government programs.

The benefits of proper planning cannot be overemphasised, even in Agribusiness. The future, its tendency of being uncertain due to factors which we may not have control over (especially in agriculture with variables such as changing climate conditions, production risk, financial risks, changes in government policies e.t.c) can to a certain extent be predicted, or a more favourable outcome assured through proper planning.

Agribusiness, like every other business, requires planning regardless of the inevitable circumstances which may likely arise. Through these, one is able to define the goal strategically and measure progress over time.

This publication discusses the importance of business planning for agricultural firms—from input suppliers to producers to processors—and describes the steps required to prepare a thorough business plan. The general process of business planning is the same for each type of firm. However, each may have differing individual aspects that affect its plan's contents. Regardless, we present a recommended format that should be useful for all types of agricultural firms as they develop written business plans. We use examples from the wide variety of agribusinesses to provide a broad context to the general theme of business planning.

In order to develop a complete agribusiness plan, it is necessary to recognize all the potential risks. The uncertainties are not limited to production, but to all facets of a business. A comprehensive manual on how to write an agriBusiness Plan was developed by Extension Agricultural Economist of Texas A&M University System. The manual gives a complete overview of the importance of agribusiness planning and how each issue should be addressed. Each topic includes a publication for reference, necessary worksheets for completing that lesson and a case study application. A farmer interested in doing an AgriBusiness Plan is advised to read and understand the manual, and then follow each step in the manual to write down his own business plan. Having a written AgriBusiness Plan has many potential uses, but overall it would help you to understand better your risks, have a plan for each one of them, and react accordingly.

The following components of an AgriBusiness Plan include:

1. Executive Summary: The purpose of the executive summary is to combine all the components of the business plan into its principal points in one summary document. It must cover the information in the report in enough detail such that the reader can quickly formulate a picture of the operation. The executive summary is written after all the parts of the business plan have been completed. Follow this link to know more about the steps to consider when writing the business plan.

2. Business Organization: The business plan begins with a summary of the pertinent information regarding the operation. This includes names and contact information of all involved in making decisions, managing risk for the operation, and/or providing counsel. It provides a combined list of all the necessary information to contact the owners, managers, and others that have an impact on the operation in one central location.

3. Business History: If you are new to farming, you might not have a Business History, which can be omitted. However, if you have farmed, as it is the case for most Small Acreage Farmers in South Texas, then you will benefit from writing it. In this component, you should briefly describe how the operation came to be and how you came to be its owner and/or manager. This would include when and how was the operation started, the location of the operation, the source of the land, equipment and other resources, and was it inherited or purchased or is it rented. How was and is the operation financed? What management changes have occurred over its history? What were the defining moments or events that caused you to control the operation?

4. Mission Statement: Every operation has a reason for taking on the risks associated with agricultural production. The purpose of the mission statement is to precisely and emphatically state why your operation exists. It should focus each person involved in the operation each day. Anyone working within the operation should, upon reading the mission statement, know how his or her daily tasks, once completed, have helped to fulfill the operations mission. The mission statement should answer three key questions:

What are the opportunities or needs that we exist to address? (The purpose of the organization)

What are we doing to address these needs? (The business of the organization)

What principles or beliefs guide our work? (The values of the organization).

5. Resource Inventory: Agricultural producers use resources such as land, labor, machinery, breeding stock, management and financial capital to produce commodities for sale. An accurate inventory of these resources is important for the evaluation of the current health of the operation

and planning as well as a preliminary method for evaluating new enterprises. The use of an up-to-date resource inventory can help to:

Provide a current overview of the operation

Complete a balance sheet

Provide a summary of collateral that can be used for a loan

Identify the status and condition of your assets and liabilities

Evaluate options for growth and diversification

6. SWOT Analysis: A SWOT analysis is a term used to describe a tool that is effective in identifying your Strengths and Weaknesses, and for examining the Opportunities and Threats you face. While it is a basic, straightforward model, it has been a popular business practice for many years because it helps provide direction and serves as a basis for the development of business plans. It accomplishes this by examining the strengths (what an operation does well) and weaknesses (what an operation does not do well) in addition to opportunities (potentially favorable conditions for an operation) and threats (potentially unfavorable conditions for an operation). Once completed, the SWOT analysis can help determine if the information indicates something that will assist the operation in achieving its objectives (a strength or opportunity), or if it indicates an obstacle that must be overcome or minimized to achieve desired results (weakness or threat).

7. Legal and Liability Issues: Contemplating the prospects of legal action against your farming or ranching operation can be unnerving. The assorted unknown variables can make the planning process frustrating, while the threat of enormous adverse judgments generates an emotional response. Planning for the loss or withdrawal of a key member of your management team can also be unpleasant. The business planning process will help identify risks and assign resources to manage these risks. A large number of potential legal and liability issues will be identified with a thorough SWOT analysis. Similarly, a resource inventory will help identify the individuals and tools necessary to manage these risks. It will be impossible to list all potential liability risks as well as to completely eliminate all of the legal and liability risk from an activity as inherently risky and uncertain as agriculture. The completed business plan will address specific action plans for dealing with identified risks, as well as outline contingencies for the unforeseen risks.

8. Setting Goals: One of the most important aspects of business planning is the setting of goals. Properly defined goals can assist the operation's management team in determining whether the operation is moving forward. Goals can be either short term or long term, however, each short term goal should correspond to a long term goal, and each of these should move the operation towards fulfilling its mission statement. Goals should be incorporated into each of the financial, production and marketing components. One common theme found in most business planning publications regarding goals is the concept of SMART goals. This concept suggests that goals must be Specific, Measurable, Attainable, Rewarding and there should be a Time-frame specified for reaching each goal.

9. Production Plan: The production plan conveys the type and quantity of commodities to be produced. The production plan pairs information from the resource inventory and financial records to serve as a realistic estimate of current activities and their anticipated financial results. A thorough production plan should detail all enterprises on an operation (crop, livestock, and other) so that scheduling of labor and financial resources can be easily examined. The production plan should provide a basis for projecting future operational activities and alternative enterprises. While changes will occur, the production plan serves to document historical performance and project the future direction of the business. Crop production plans should include the estimated acreage and yield for each crop. Estimated production levels can then be combined with anticipated prices to generate some of the figures needed for the financial component. The livestock production plan must clearly identify all related production information, including the size of the herd, cull rates, weaning rates, weaning weights, rates of gain, purchase price, sales prices, etc. In addition, details regarding the replacement herd and breeding herd should be described.

10. Financial Plan: The financial plan component of the business plan serves as the heart of the overall plan and has three main objectives. The first is to identify where the operation is financially (Financial Position). The second objective is to determine how the operation performed during the previous year (Financial Performance). Finally, the third objective is to provide an analysis of where the operation will be in the future (Financial Projections). These three objectives are discussed further in this link.

11. Marketing Plan: It is essential for an agricultural producer to have a written marketing plan. Developing a good marketing plan will help you identify and quantify costs, set price goals, determine potential price outlook, examine production and price risk, and develop a strategy for marketing your crop. While producers have traditionally done a good job of producing, they have often neglected marketing. In the past, farm loan programs and deficiency payments allowed producers to neglect or ignore the marketing side of their businesses. Now, with the possible elimination of farm programs and increased volatility in the markets, producers will have the right and the obligation to determine their own financial security. In this more uncertain and risky future, failing to plan may be the same as planning to fail.

12. Example AgriBusiness Plan: When you start writing you Agribusiness Plan, follow the example provided in the manual to guide your work.

About the Business Plan

One of the most important documents for any business is their business plan. It is common practice for consultants, lenders, potential business partners, and other business-associated individuals to request a business plan to make a more informed decision concerning their relationship with a business. However, business plans have many more direct benefits for the business owner. The planning process forces owners to systematically consider all facets of the business. In so doing, they become more knowledgeable of the business, the industry, and the market environment in which their business operates. The process also helps to define business goals and to assess the impact that uncertainty may have on future business outcomes. Perhaps most importantly, the written plan provides a well-defined direction for the business. Therefore, it can be used to keep all employees moving toward the common goals established within it.

Completing a business plan can be a time-consuming activity, but well worth the effort. Because businesses operate in an ever-changing environment, the plan should be revisited periodically to be sure that the business is headed in the proper direction or to formally alter the firm's course if circumstances dictate that this is necessary. Again, the systematic review of the business plan forces the owner, and potentially others, to look at the business as a whole and make better-informed decisions.

We provide an example format for you to use as a guide in developing your plan. Notice that there are several topics that should be addressed, corresponding to the four functional areas of management: marketing, production, finance, and human resources. By developing a section for each of these topics, the plan will be easy to follow as you revisit it or as others review it. You should take some liberty as you develop your plan; feel free to customize it in a way that will fit your specific circumstances.

Forecasting in Agribusiness

Appropriate forecasting is an extremely important factor for the success of any business related to the Food and Fiber Industry. Forecasting with the right techniques is needed to provide with a correct assessment for decision-making purposes. Time-Series Analysis is commonly used since most of the data sets we study in Agricultural Economics are observed over time, where the order of observation is an integral part of the analysis. So, for instance, data on prices in efficient markets are such that prices observed in close time proximity to one another are closer to each other than are prices observed at long time intervals. Data from such markets will show no particular devotion to their historical mean. Many standard (non-time series) statistical and econometric methods rely on an informative mean for their desirable properties. Failure to correctly account for the time series properties of one's data may result in the fitting of spurious relations between or among economic data sets.

Risk and uncertainty are central to forecasting and prediction; it is generally considered good practice to indicate the degree of uncertainty attaching to forecasts. The process of climate change and increasing energy prices has led to the usage of Egain Forecasting of buildings. The method uses Forecasting to reduce the energy needed to heat the building, thus reducing the emission of greenhouse gases.

Practice of Customer Demand Planning

Forecasting is used in the practice of Customer Demand Planning in every day business forecasting for manufacturing companies. The discipline of demand planning, also sometimes referred to as supply chain forecasting, embraces both statistical forecasting and a consensus process. An important, albeit often ignored aspect of forecasting, is the relationship it holds with planning. Forecasting can be described as predicting what the future will look like, whereas planning predicts what the future should look like.

There is no single right forecasting method to use. Selection of a method should be based on your objectives and your conditions (data etc.). A good place to find a method, is by visiting a selection tree. An example of a selection tree can be found here.

While the term "forecasting" may appear to be rather technical, planning for the future is a critical aspect of managing any organization—business, nonprofit, or other.

In fact, the long-term success of any organization is closely tied to how well the management of the organization is able to foresee its future and to develop appropriate strategies to deal with likely future scenarios. Intuition, good judgment, and an awareness of how well the economy is doing may give the manager of a business firm a rough idea (or "feeling") of what is likely to happen in the future.

Nevertheless, it is not easy to convert a feeling about the future into a precise and useful number, such as next year's sales volume or the raw material cost per unit of output. Forecasting methods can help estimate many such future aspects of a business operation.

Suppose that a forecast expert has been asked to provide estimates of the sales volume for a particular product for the next four quarters. One can easily see that a number of other decisions will be affected by the forecasts or estimates of sales volumes provided by the forecaster. Clearly, production schedules, raw material purchasing plans, policies regarding inventories, and sales quotas will be affected by such forecasts. As a result, poor forecasts or estimates may lead to poor planning and thus result in increased costs to the business.

How should one go about preparing the quarterly sales volume forecasts? One will certainly want to review the actual sales data for the product in question for past periods. Suppose that the forecaster has access to actual sales data for each quarter over the 25 year period the firm has been in business. Using these historical data, the forecaster can identify the general level of sales. He or she can also determine whether there is a pattern or trend, such as an increase or decrease in sales volume over time. A further review of the data may reveal some type of seasonal pattern, such as peak sales occurring before a holiday.

Thus by reviewing historical data over time, the forecaster can often develop a good understanding of the previous pattern of sales. Understanding such a pattern can often lead to better forecasts of future sales of the product.

In addition, if the forecaster is able to identify the factors that influence sales, historical data on these factors (or variables) can also be used to generate forecasts of future sales volumes.

Forecasting Methods

All forecasting methods can be divided into two broad categories: qualitative and quantitative. Many forecasting techniques use past or historical data in the form of time series. A time series is simply a set of observations measured at successive points in time or over successive periods of time. Forecasts essentially provide future values of the time series on a specific variable such as sales volume. Division of forecasting methods into qualitative and quantitative categories is based on the availability of historical time series data.

Qualitative Forecasting Methods

Qualitative forecasting techniques generally employ the judgment of experts in the appropriate field to generate forecasts. A key advantage of these procedures is that they can be applied in situations where historical data are simply not available. Moreover, even when historical data are available, significant changes in environmental conditions affecting the relevant time series may make the use of past data irrelevant and questionable in forecasting future values of the time series. Consider, for example, that historical data on gasoline sales are available.

If the government then implemented a gasoline rationing program, changing the way gasoline is sold, one would have to question the validity of a gasoline sales forecast based on the past data. Qualitative forecasting methods offer a way to generate forecasts in such cases. Three important qualitative forecasting methods are: the Delphi technique, scenario writing, and the subject approach.

Delphi Technique

In the Delphi technique, an attempt is made to develop forecasts through "group consensus." Usually, a panel of experts is asked to respond to a series of questionnaires. The experts, physically separated from and unknown to each other, are asked to respond to an initial questionnaire (a set of questions). Then, a second questionnaire is prepared incorporating information and opinions of the whole group. Each expert is asked to reconsider and to revise his or her initial response to the questions.

This process is continued until some degree of consensus among experts is reached. It should be noted that the objective of the Delphi technique is not to produce a single answer at the end. Instead, it attempts to produce a relatively narrow spread of opinions—the range in which opinions of the majority of experts lie.

Scenario Writing

Under this approach, the forecaster starts with different sets of assumptions. For each set of assumptions, a likely scenario of the business outcome is charted out.

Thus, the forecaster would be able to generate many different future scenarios (corresponding to the different sets of assumptions). The decision-maker or businessperson is presented with the different scenarios, and has to decide which scenario is most likely to prevail.

Subjective Approach

The subjective approach allows individuals participating in the forecasting decision to arrive at a forecast based on their subjective feelings and ideas. This approach is based on the premise that a human mind can arrive at a decision based on factors that are often very difficult to quantify.

"Brainstorming sessions" are frequently used as a way to develop new ideas or to solve complex problems.

In loosely organized sessions, participants feel free from peer pressure and, more importantly, can express their views and ideas without fear of criticism. Many corporations in the United States have started to increasingly use the subjective approach.

Quantitative Forecasting

Quantitative forecasting methods are used when historical data on variables of interest are available—these methods are based on an analysis of historical data concerning the time series of the specific variable of interest and possibly other related time series. There are two major categories of quantitative forecasting methods.

The first type uses the past trend of a particular variable to base the future forecast of the variable. As this category of forecasting methods simply uses time series on past data of the variable that is being forecasted, these techniques are called time series methods.

The second category of quantitative forecasting techniques also uses historical data. But in forecasting future values of a variable, the forecaster examines the cause-and-effect relationships of the variable with other relevant variables such as the level of consumer confidence, changes in consumers' disposable incomes, the interest rate at which consumers can finance their spending through borrowing, and the state of the economy represented by such variables as the unemployment rate.

Thus, this category of forecasting techniques uses past time series on many relevant variables to produce the forecast for the variable of interest. Forecasting techniques falling under this category are called causal methods, as the basis of such forecasting is the cause-and-effect relationship between the variable forecasted and other time series selected to help in generating the forecasts.

UNDERSTANDING OF BEHAVIOUR OF TIME SERIES

Before discussing time series methods, it is helpful to understand the behavior of time series in general terms. Time series are comprised of four separate components: trend component, cyclical component, seasonal component, and irregular component. These four components are viewed as providing specific values for the time series when combined.

In a time series, measurements are taken at successive points or over successive periods. The measurements may be taken every hour, day, week, month, or year, or at any other regular or irregular interval. While most time series data generally display some random fluctuations, the time series may still show gradual shifts to relatively higher or lower values over an extended period. The gradual shifting of the time series is often referred to by professional forecasters as the trend in the time series. A trend emerges due to one or more long-term factors, such as changes in population size, changes in the demographic characteristics of population, and changes in tastes and preferences of consumers. For example, manufacturers of automobiles in the United States may see that there are substantial variations in automobile sales from one month to the next. But, in reviewing auto sales over the past fifteen to twenty years, the automobile manufacturers may discover a gradual increase in annual sales volume.

In this case, the trend for auto sales is increasing over time. In another example, the trend may be decreasing over time. Professional forecasters often describe an increasing trend by an upward sloping straight line and a decreasing trend by a downward sloping straight line. Using a straight line to represent a trend, however, is a mere simplification—in many situations, nonlinear trends may more accurately represent the true trend in the time series.

Although, a time series may often exhibit a trend over a long period, it may also display alternating sequences of points that lie above and below the trend line. Any recurring sequence of points above and below the trend line that last more than a year is considered to constitute the cyclical component of the time series—that is, these observations in the time series deviate from the trend due to cyclical fluctuations (fluctuations that repeat at intervals of more than one year). The time series of the aggregate output in the economy (called the real gross domestic product) provides a good example of a time series that displays cyclical behavior. While the trend line for gross domestic product (GDP) is upward sloping, the output growth displays a cyclical behavior around the trend line. This cyclical behavior of GDP has been dubbed business cycles by economists.

The seasonal component is similar to the cyclical component in that they both refer to some regular fluctuations in a time series. There is one key difference, however. While cyclical components of a time series are identified by analyzing multiyear movements in historical data, seasonal components capture the regular pattern of variability in the time series within one-year periods. Many economic variables display seasonal patterns. For example, manufacturers of swimming pools experience low sales in fall and winter months, but they witness peak sales of swimming pools during spring and summer months.

Manufacturers of snow removal equipment, on the other hand, experience the exactly opposite yearly sales pattern. The component of the time series that captures the variability in the data due to seasonal fluctuations is called the seasonal component.

The irregular component of the time series represents the residual left in an observation of the time series once the effects due to trend, cyclical, and seasonal components are extracted. Trend, cyclical, and seasonal components are considered to account for systematic variations in the time series. ‘h e irregular component thus accounts for the random variability in the time series. The random variations in the time series are, in turn, caused by short-term, unanticipated and nonrecurring factors that affect the time series. The irregular component of the time series, by nature, cannot be predicted in advance.

Planning in Agribusiness the Importance

Here are some of the things that proper planning helps us achieve at Babban Gona.

Increases efficiency

Helps in achieving goals

Planning provides focus and aids coordination

It aids decision making

It reduces risks and oversight

Planning encourages innovation and creativity

It aids delegation

Planning in agribusiness Increases efficiency

For us at Babban Gona, planning helps to increase the efficiency of our business. We are able to determine per time what to invest in and how much resources we throw into specific projects. For every planting season, our farmers benefit from the end-to-end suite of services we provide from the beginning of the planting season to the very end (down to the bags, threads, and needles needed at harvest). The resultant effect of this is a reduction in waste and that we effectively manage our resources to ensure effectiveness and productivity.

Helps in achieving goals

Through planning, we are able to set specific, practical steps that enable the achievement of objectives and goals. It reminds us of our objectives and brings about clarity. Some of our aspirations as an organisation are to create more jobs and ensure that we contribute our quota in making Nigeria attain food security which would then translate into an improved economy. Our services as an organisation are tailored towards this end and through planning, we take measured steps based on how it affects our big picture.

Planning provides focus and aids coordination

It provides focus for everyone including the personnel, everyone is able to prioritise and maximise resources available to ensure that each task is attended to at the right time. It also ensures coordination and helps everyone at Babban Gona stay in sync with each other.

It aids decision making

Through planning, we are equipped to make the best decision when faced with different options, it reduces the chances of acting based on impulse. For example, knowing what sort of innovations to invest in per time based on the need of our business. This also includes building strategic partnerships and collaborating with the right organisations, so far, planning has been instrumental for us in making decisions as an organisation.

It reduces risks and oversight

It reduces the risks and chances of losing rather than making profits due to an oversight, for example, at Babban Gona we have been able to maintain repayment rates of up to 99% on loans granted through the strong risk mitigations systems which we have put in place. Thereby, ensuring both the income of the farmers and seeing that we are not at a loss as well. Planning basically involves envisaging all possible uncertainties and working to reduce the damage.

Planning in Agriculture- Artificial Intelligence

Planning Encourages Innovation and Creativity

At Babban Gona we encourage innovation and creativity, we invest in great minds and talents who come up with ideas to make our processes better. These ideas reflect in our adoption of Artificial Intelligence in dealing with our farmers, in financing- making it easier for our farmers to have access to cost-effective credit, in training and development- to improve methods and ensure the effectiveness and productivity of our farmers.

Planning in Agribusiness Aids Delegation

Our team leads and managers easily delegate authority knowing that the direction is clear. The advantage of this is not having to be present at all times and being able to focus on other aspects of the business.

However, plans do not have to be rigid, they can always be changed and amended based on the physical realities of the business, businesses operate in an ever-changing environment and for one to remain relevant, your business should be positioned to move along the current trend and reality. Investors and financial institutions also would feel safe knowing that their investments are put to good use, they are accounted for and will yield returns. This can only be achieved and tracked through proper planning and implementation.

10

PRODUCTION AND INVENTORY MANAGEMENT IN AGRIBUSINESS

A solution for agriculture inventory management delivers full operational visibility and control to agribusinesses that supply farmers and growers. With the agriculture industry contributing over a trillion dollars to the U.S. annual gross domestic product, it is essential for manufacturers and distributors of farm machinery, seeds, agrichemical products, animal feed, plants, and produce to be operationally strong within the supply chain.

Acctivate agriculture inventory management software provides tools to manage a business' entire workflow from procurement to sales to fulfillment, all while finances are synced from QuickBooks through a bi-directional integration.

Many agribusiness firms do not actively manage inventory. This does not mean that they ignore inventory. Rather, they hold large inventories because any potential savings from inventory reductions are far outweighed by the inventory-induced reductions in production, procurement, or transportation costs. Often economies of size cause long productions runs which lead to inventory accumulation. Simultaneously, seasonality leads to inventory buildups of key inputs like seed as well as outputs like corn. Economies in procurement such as forward buying in the food industry and quantity discounts increase inventories. Similarly, unit trains and other forms of bulk shipping discounts contribute to inventory buildups.

Many argue that the focus point (and perhaps the linchpin) of successful supply chain management is inventories and inventory control. So how do food and agribusiness companies manage their inventories? What factors drive inventory costs? When might it make sense to keep larger inventories? Why were food companies quicker to pursue inventory reduction strategies than agribusiness firms?

In 1992, some food manufacturers and grocers formed Efficient Consumer Response to shift their focus from controlling logistical costs to examining supply chains. Customer service also became a key competitive differentiation point for companies focused on value creation for end consumers. In such an environment, firms hold inventory for two main reasons, to reduce costs and to improve customer service. The motivation for each differs as firms balance the problem of having too much inventory (which can lead to high costs) versus having too little inventory which can lead to lost sales.

AGRICULTURE INVENTORY MANAGEMENT SOFTWARE FOR MANUFACTURING

Whether an agribusiness produces products using food ingredients, chemical formulas, or parts for equipment, agriculture inventory management with Acctivate helps them oversee the manufacturing process. All production processes are enhanced with multi-level bill of materials to specify components/ingredients, workflow management capabilities, product lot and serial number traceability, and more to deliver superior visibility.

Manage formulas or recipes with the ability to adjust accordingly for ingredient and yield variations.

Build products with components/parts as kits ordered on demand or assemblies made for stocking.

Accommodate kit and assembly customizations when there are one-off adjustments to component quantities or component removals/additions.

Solid agriculture inventory management is essential for agribusinesses

Small to midsized agribusinesses can remain viable within the supply chain with Acctivate as a complete agriculture inventory management solution. The simple, but powerful capabilities of Acctivate enable SMBs to match or one-up larger organizations as all operations are consolidated and automated within the one system.

Production Management

Almost seventy percent of the world's agricultural food production particularly, in crops is produced by a considerable number of smallholders especially in less developed economies. A considering number of these smallholder farmers live in abject poverty, have significantly low productivity, resulting in low incomes and resort to poor methodology in farming that deplete the soil. Forest in the tropic have been destroyed to gratify the need for fertile farm lands, therefore it is imperative that we employ methods that will be sustainable in the long run. Blue chip companies on the globe are conscious of the need to produce sustainably and have shown the commitment towards that. It is in view of this that, Solidaridad and a number of other Non Governmental Organization's as well as multi National Companies decided on the Utz Certified standard for sustainable cocoa in 2009. "At the close of 2009 close to 3,600 smallholders cocoa producers had been trained in certification standards to produce sustainably.

Supplying enough food for the mushrooming global population in environmentally and sustainably tolerable levels, devoid of negotiating on our resources, food production should optimize tremendously from the present production. It should strongly be noted that we can maximize productivity without necessarily damaging the ecology to meet the growing food needs of our time and future. According to the Institute of Food Technologist, (2010 Pg. 574), during the global summit on food security in 2009, it was asserted that, food production must grow between 34%-70% by 2050, if we are to feed the estimated 9 billion people. The obstacle however is the big, increasing food security need in most economies of the world. A large proportion of food produced in less developed and developing economies are not consumed, partly due to poor management and distribution network. In view of this, pronounce malnourishment and dietary deficit in vitamins, minerals, protein, and calories continuous to be common globally. To overturn this dreadful development and ensure a momentous sustainable improvement towards improving the impoverished, this theme should attract the attention of managers of our economies and international bodies

Low cost scientific improvements in food production and distribution mechanisms as well as environmental management and dealing decisively with post harvest management will be key in satisfying this need, indeed failure on our part to hasten the application of science today and yet to be discovered; to food production in an ecologically tolerable level will leave us in critical condition. In fact it will be a fiasco if we increase food production without improving conservation. Variables contributing to food losses differ from one economy to the other as a result of differentials in infrastructure, investments and awareness levels

The milestone of this paper will concentrate on agriculture primary food production since food production encompasses a wide range of produce, thus all issues will be constricted to improving farm produce to meet the growing demand of the global market, whiles considering the sustainability of such produce with the view to ensuring exponential increase in farmers' incomes.

There are four over-arching goals in writing this paper. The first, and most basic, is discussing challenges confronting increased sustainable agriculture food production The second is discussing the underpinning sustainability by restoring and conserving the natural resource base, managing the ecological and societal sustainability issues connected to the development of food crop production by optimizing performance in existing farms and improving the income levels of smallholder producers.

The third objective is to discuss hopeful but innovative channels of forward looking approach by presenting a promising way of increasing food production sustainably. The spotlight of this essay will begin with a prologue that abridge the growth of cultivation, food knowledge and sustainability perspective; a section on challenges confronting optimum crop production to meet the needs of humanity; a section about potential environmental and social consideration solutions; a section on forward looking approach and concluding remarks.

Limitations of Increased And Sustainable Agricultural Food Production

"World population is expected to rise by 36% between 2000 and 2030. Average crop yields are anticipated to grow at about the same rate. However, relative yield increases have been undermined generally (projections for global yields in the next decade predict 1-1.1% p.a. growth for cereals)."

The reasons are farfetched; weak agronomic practices in the form of poor: production, maintenance of farm, and post harvest management as well as international charters have been the bane for most smallholder producers in sub-Saharan Africa.

With respect to production, most often than not, the soil is not prepared and where it is done, the preparation is woefully executed, "Africa is the lone constituency on the globe where per capita food productivity has been waning for the past three decades. Cereal yields in Africa are a quarter of the international average and Africa's soils remain the worst globally." Karapinar, Baris Haberli Christian (2010, Pg. 82); most smallholder farmers adopt low yielding planting materials, and where they are used, they are not appropriate for the soils and climate; more imperatively, depleted soil fertility; absence of small irrigation systems due to over reliance on rain, Crop plants require a continuous supply of water to replace the water evaporated (transpired) from their aerial organs; irresponsible use of agro-chemicals; non observance of proper planting spacing that are necessary within the interrows immensely impinge on harvesting paths, weed control, fertilization and other agronomic practices; the high planting density in planting pattern leads to competition for basic crop requirements.

On the other hand, a sparse planting density (evident in some farms) is a waste of land resources. For good growth and productivity, the optimum spacing for every food crop farm should be observed to circumvent crop plant from being tall and etiolated.The constant attack of insect, pests and diseases on farms all contract yields; ill timed harvesting periods are equally contributing variable to low productivity, thus long and short harvesting cycle if not appropriate to the food crop affect the quality of the produce to a larger extent.

International bodies such as the World Bank, IMF and other powerful bodies have via policies contributed to low productivity unconsciously, the inexorable of such policies and financial support to developing economies is migration of needed farm labor in food producing areas to the urban slums. Regarding farm maintenance, "the maintenance regime (weeding, pruning) on most farms are generally poor. There were cases of excessive pruning which tended to affect the architecture of the palms. An instance of fronds supporting developing bunches being cut was observed. None of the farms had been planted with cover crops and there were no harvesting paths" CSIR-OPRI, (2012). The figures below are illustration of poorly maintained oil palm farm at Bonsaaso Ghana

Supply Chain Management

A common perception and experience is that supply chain management leads to cost savings, largely through reductions in inventory. Inventory costs have fallen by about 60% since 1982, while transportation costs have fallen by 20%. Such cost savings have led many to pursue inventory-reduction strategies in the supply chain. To develop the most effective logistical strategy, a firm must understand the nature of product demand, inventory costs, and supply chain capabilities.

Firms use one of three general approaches to manage inventory. First, most retailers use an inventory control approach, monitoring inventory levels by item. Second, manufacturers are typically more concerned with production scheduling and use flow management to manage inventories. Third, a number of firms (for the most part those processing raw materials or in extractive industries) do not actively manage inventory.

Yet, such firms must be alert to changing conditions that may require more exact inventory management. One example would be if crops are marketed as small lots of value-added grain instead of commodities. Production proliferation in the seed industry may be another instance. Finally, whether due to food safety concerns, GMOs, food labeling, or the growth of organic food markets, identity preservation requires more precise inventory control.

Inventory Management

Inventory management is influenced by the nature of demand, including whether demand is derived or independent. A derived demand arises from the production of another product. For example, when John Deere knows its demand for a tractor, it can simply compute the demands for the parts, materials, and components needed to produce that tractor. Manufacturers of all sizes use such calculations which are part of flow management to manage inventories, schedule deliveries for inputs, and manage capacity. Flow management software has evolved from Materials Requirements Planning (or MRP) in the 1960s to the much more complex Enterprise Resource Planning (or ERP) of the 1990s. A flow management system is set in motion by the demand for end products.

Independent demand arises from demand for an end product. End products are found throughout a supply chain. Wheat is an end product for a grain elevator, as is flour for a miller or cereal for a grocer. By definition, an independent demand is uncertain, meaning that extra units or safety stock must be carried to guard against stockouts. Managing this uncertainty is the key to reducing inventory levels and meeting customer expectations. Supply chain coordination can decrease the uncertainty of intermediate product demand, thereby reducing inventory costs.

Customer Service and Inventory

The availability of inventory provides customer service. The Item Fill Rate (IFR) measures how often a particular product (often called a stock keeping unit or SKU) is available. A common metric of customer service, IFR is expressed as the percentage of time that a customer can obtain the item they seek. A firm may set its customer service order policy at 95%, seeking to fill 95% of the orders for an item from inventory.

However, life is a bit more complicated. A customer might not obtain what they seek for several reasons. The seller may have run out of a product due to an inaccurate forecast. Or the supplier may have shipped an incorrect package size or flavor. Products in inventory may be unfit for sale because of damage or an expired shelf life. Finally, a seller may not have the capability to accurately track inventory in their stores or distribution centers.

Safety Stock

To avoid shortfalls or stockouts, firms carry extra inventory known as safety stock. As more customer service is provided, a firm can expect sales to increase. However, as a firm tries to provide perfect customer service, logistical costs increase exponentially. Also, if a firm holds too much inventory, it can lead to low inventory turnover and hide operational problems. For example, carrying too much stock means that you might not discover that your supplier is frequently late with delivery times.

The structure of independent demand and logistical requirements vary by stage in the product life cycle (introduction, growth, maturity, and decline). During introduction, logistics must support the business plan for product launch, while preparing to handle potential rapid growth by quickly expanding distribution. At market maturity, the logistical emphasis shifts to become cost driven. In the decline stage, cash management, inventory control, and abandonment timing become critical. Over-abundance of products in the late maturity or decline stage will eventually result in obsolete products. The obvious difficulty is predicting how long each stage will last and how abruptly sales will fall in the decline stage.

The life cycle strategy typically involves getting to profitability quickly recuperating startup costs, then sustaining high profits for as long as possible, and finally acting decisively for products in decline to minimize losses. Understanding this life cycle can help managers select logistical tactics, inventory levels and supply chain designs. The ultimate goal for companies should be to have just enough inventory to satisfy consumer demand.

Another life cycle attribute is that demand uncertainty shifts as we progress through time. Product managers face substantial uncertainty during the introduction and growth stages, relative stability during maturity, and increasing uncertainty in decline. This uncertainty drives forecasting accuracy and the level of safety stock required to meet customer service expectations.

The coefficient of variation (CV) measures the stability of a product's demand, comparing the variability in demand to the size of the average demand. High demand variability in the introductory stage means it is difficult, if not impossible, to forecast demand. Thus, high levels of inventory must be held to meet even minimal customer service levels. In contrast, lower variability during maturity means that demand forecasts are quite accurate. However, inventory levels may still be large because they are based on larger sales volumes.

In addition to the vagaries associated with product life cycle stage, two other sources of uncertainty also drive the level of inventory. First, demand can vary from day to day, week to week, or seasonally. Second, there may be variability in lead time, or the time from when an order is placed until delivery is made.

Forecasting demand used to be more exact because products stayed in the mature product life cycle phase for a long time. Today many companies find it far more difficult to forecast sales because of product proliferation. Product line extensions result in more products that cannibalize sales and shorten the life cycle. Thus, more sales are coming from products in the erratic earlier stages of life, as opposed to sales from products in the mature stage of the life cycle.

Inventory Costs

Different models are used to manage inventory for products that are continually available like milk or products available for limited time like seed. The Economic Order Quantity (EOQ) model determines the least cost level of inventory to carry, as well as costs. News Vendor models are used for products only available for a single period.

EOQ and News Vendor models have proved useful for managing inventory for many years, analyzing tradeoffs among major cost components. These models are robust and easy to customize to particular industries. Their approach to costing is similar reflecting levels of inventory, as well as shipping costs or quantity discounts.

Inventory costs fall into three classes: (a) carrying costs of regular inventory and safety stock; (b) ordering or setup costs; and (c) stockout costs. Inventory control systems balance the cost of carrying inventory against the costs associated with ordering or shortfalls.

First, carrying cost (or a cost to hold inventory) is comprised of capital costs, service costs, storage costs, and risk costs. A carrying cost involves the opportunity cost for holding inventory. If the firm did not have money tied up in inventory, it could either use the savings to make investments in other assets or pay down debt. Thus, a firm should first determine what it would do with any savings from a reduction in inventory. If the dollars are used to buy capital equipment, an appropriate opportunity cost is the firm's hurdle rate or its "required rate of return." If the dollars are used to pay down debt, the interest rate on the loan should be used to value the inventory. The other three aspects of carrying cost are non-capital costs.

The service costs are often masked in a firm's fixed costs. A firm should determine how much of its insurance and tax expense is associated with inventory. This is especially important in states that have an inventory tax. A firm has cash outlays for warehouses and materials handling equipment, either owning or leasing space from a distributor.

In either case, the firm should determine how much is spent on space. Inventory risk reflects characteristics of the product. Some items are more prone to be stolen, others are more likely to be damaged, yet others may become obsolete before a sale is made. In any case, risk means that if too much inventory is held, a certain proportion of the inventory will be unavailable for production or sale.

INVENTORY REDUCTIONS

To determine the cost of carrying inventory, one needs to know the average quantity of inventory, an inventory carrying cost (as a percent of product cost), and the average cost per unit of inventory. If a firm plans to use inventory reductions to fund other capital assets, inventory carrying cost might be 30% (25% for an opportunity cost and 5% for the service, space, and risk costs). If the firm plans to use the savings to reduce debt, the appropriate rate might be 12% (7% for the interest rate and 5% for the other costs). Regardless of the carrying cost rate being used, as a firm holds more inventory, carrying cost increases.

Firms carry extra inventory to guard against uncertain events. Known as safety stock, the purpose of this inventory is to provide protection against stockouts. Safety stock is costed just like regular inventory, it is an interest rate times the level of safety stock. The level of safety stock required to guard against a stockout depends upon the customer service level, the standard deviation of demand of the product, and lead time. Let's explain in greater detail.

Assume that it takes ten days from the time an order is placed until a shipment arrives and that on an average 20 cases are sold each day. Thus, over the teb days that we are waiting for the delivery (our lead time), we expect to sell 200 cases. If we trusted our forecast, supplier, and trucking company, we would simply hold 200 cases for the ten days. But we realize that forecasts are inaccurate, some suppliers are unreliable, and shipping times vary. If less is sold than expected during the ten days or if the shipment arrives early, we will still have inventory on the ten day and no customer service problems are encountered. However, if sales are above expectations during the ten days or deliveries are late, we might run out (or stockout) of product.

Managing the uncertainty surrounding safety stock is the key to reducing inventory levels. But in today's competitive environment, it is difficult to lower safety stock requirements for two reasons. First, some buyers especially large retailers are requiring higher customer service levels, which raise safety stock levels. Second, the product mix for many firms includes more new products with the corresponding greater demand variability. Thus, most firms seeking to reduce safety stock can only do so by focusing on aggressively cutting lead times.

The second cost to consider is ordering costs. Ordering costs include a cost for transmitting the order, receiving the product and placing it into storage, inbound transportation, and processing the invoice. Recent advancements in information technology have lowered this cost by a factor of six for many industries. A manufacturer uses the cost of a production setup instead of an ordering cost.

Finally, stockout costs involve lost sales when no inventory is on hand. Such costs fall as inventory (and customer service) levels increase. The relationship between stockout costs and inventory depends upon the accuracy of the demand forecast and the ability of the firm to recognize and react to a change in demand. Stockout costs depend on how a customer reacts to a stockout, the frequency of stockouts, and the availability of substitute products. Stockout costs can be very high if a lack of substitute products means that a customer will switch suppliers. In contrast, if buyers simply substitute a different product, stockout costs may be inconsequential.

In practice, many firms do not assess stockout costs because different divisions of a firm cannot reach agreement on what is the cost of running out. Marketing may desire a very high stockout cost to force a penalty cost on running out. Operations or finance may resist this as it leads to inventory buildups.

Service level goals can differ by the value placed on stockouts and indirectly carrying costs. A high stockout valuation will result in higher inventories and higher service levels. One way to evaluate an inventory management policy is to choose a service level target. From this target, the inventory policy will determine the inventory requirements and associated costs of providing that level of service. A higher service level implies that more inventory will be held as safety stock. The tradeoff decision occurs at the point where the cost of carrying extra safety stock balances the stockout cost.

Inventory levels are affected by customer service expectations, demand uncertainty, and the flexibility of the supply chain. For products with relatively certain demand and a long product life, it should be relatively easy to maintain desirable customer service standards even as inventories are reduced. However, for products characterized by erratic demand, a short life cycle, or product proliferation, a more responsive supply chain and larger buffer inventories may be needed to meet a desired customer service level.

Consumers are demanding more customer service from firms throughout the supply chain. Firms with high customer service levels may gain a competitive advantage over those that do not have the supply chain capabilities in place or the ability to manage them.

Firms who understand their demand recognize stockout costs and carry appropriate levels of inventory are ultimately better able to effectively manage inventory and provide the desired service level to customers. As industrialization affects agribusiness and agriculture in general, the importance of customer service and competitiveness will become critical for firms and supply chains.

The Importance of Inventory Management

Running a successful agribusiness is not easy. To achieve profitable and successful production, farmers have to be skilled in various aspects of farm management. In other words, they have to be successful managers, accountants, consultants, and scientists who are very good at performing all types of field activities. Luckily, we live in the era of modern agriculture in which farmers can rely on smart AgTech solutions to keep accurate records and manage their farm operations. Despite advancements made in AgTech software, one thing can significantly determine the success of crop management and farm operations – that's inventory management.

Inventory Management In Farm Management Software

Inventory – in simple words it means keeping inventory levels, raw materials and other products accounted for and in stock. For example, agricultural inventory, includes various fertilizers, pesticides, fuel, or seeds that farmers use in their production.

Although it doesn't sound like a big deal, healthy farm inventory management is one of the most important habits of a successful small business. Multiple aspects of agricultural production are closely related to inventory management. These are activity planning and execution, activity record keeping, ensuring complete production traceability, and planning seasonal budget.

Poor Inventory Management Leads to Cash Flop

Occupied with their daily field operations, farmers often forget that healthy inventory management is a very important aspect of their production. The consequences of poor inventory management can be significant to both farm productivity and profitability. The most common example of how inventory management impacts the success of crop production is at the time of crop sowing.

Every farmer knows that most crops, especially arable, need to be sown or planted at a specific time of the year. In order to provide the best conditions for crop growth, sowing needs to be performed during that specific period. If growers start sowing wheat and then suddenly realize that they are running low on seed stock, they need to pause sowing and make the purchase. Meanwhile, certain unfavorable conditions may occur. There is a huge possibility of a rain period and lower temperatures that can greatly impact the germination as well as the final yield.

Another common example of poor inventory management is during the harvest period. Without insight into the real-time inventory stock levels, farmers can't prepare an adequate quantity of bags, boxes, or pallets for fruit storage during the harvest. The time needed to make a purchase of stocks prolongs the time of harvest, thus affecting the fruit quality and final yield.

These two examples mentioned above are just a drop in the ocean of similar situations of improper farm inventory management.

Still Don't Have a Suitable Inventory Management Software?

Growing crops means that there are certain farm activities constantly in progress. However, many of them require tracking inventory stock levels. In other words, farmers have to be completely prepared for each activity. Otherwise, every lack of stock may lead to an additional delay in performing activities. Having the inventory software system capable of automation goes a long way in ensuring that doesn't happen.

The Economics of Farm Inventory Management

A smoothly functioning inventory system is one of the cornerstones of a successful farm operation. As the University of Minnesota notes, being able to anticipate and plan for the timely reprovisioning of inputs and materials not only reaps monetary benefits -- as in obtaining discounted prices on seed -- but it allows farmers to conduct business on-schedule and without interruption, for an added indirect economic bonus.

Types of Farm Inventory

All farm inventories are not created equal, and as the Louisiana State University Ag Center designates, different types of inventory should be recorded separately. Land itself should be inventoried distinct from machinery and equipment. The latter, however, may be tallied similarly to real estate improvements and purchased livestock. Raised livestock should follow its own inventory protocol, apart from that of market livestock. To inventory market livestock, supplies, stored crops and growing crops, provide a description, quantity, price per unit and total value per item.

Inventory Costs

One of the key economic factors that a farm manager must take into account when doing inventory is the cost of holding inventory. Washington State University and the U.S. Department of Agriculture say that for agricultural supply enterprises, the inventory holding cost should be between 20 and 35 percent of total inventory. The farm manager should adjust inventory costs to maintain that holding level. Inventory costs include storage space costs, handling costs, risk and service costs, as well as capital costs, involving interest on inventory investments.

Inventorying Methods

General lean management business principles can be applied to farm inventories to eliminate waste and avoid ordering too many supplies. The University of Minnesota discusses several inventory methods available to farm managers. Using the ABC system of inventory planning, a farmer divides his inventory into three categories: high-monetary volume items, medium- and low-monetary volume inventory. He handles high-monetary supplies one way, and follows a simpler system for lower-monetary volume supplies. For large and/or complicated orders, the economic order quantity system helps determine the optimal order size based on input cost, storage cost, order cost and annual usage.

Profitable Inventory

One of the aims of a good farm inventory system is to zero in on the most efficient and profitable amount of inventory with which to operate the farm. As WSU points out, another hidden cost to inventories is that by not having inventory in stock, the farm risks potentially losing a sale. The farmer has to balance on the one hand the upkeep costs of inventory with the sales losses that could result due to lack of inventory. To do this, he must find the breakeven point in inventory reductions, where the cost saved by eliminating inventory is outweighed by sales losses.

BIBLIOGRAPHY

- Abbott, John Cave; Food and Agriculture Organization of the United Nations (1986). Marketing Improvement in the Developing World: What Happens and what We Have Learned. Food & Agriculture Org. pp. 3–. ISBN 978-92-5-101427-1.
- Adamowicz, Mieczyslaw (2020). Bioeconomy As a Concept for The Development of Agriculture and Agribusiness. Problems of Agricultural Economics. 365: 135–155. doi:10.30858/zer/131842. ISSN 0044-1600. S2CID 234433508.
- Agricultural businesses: Key influences on growth and performance, in Agricultural Businesses: Their Growth & Performance, ISR/Google Books, 2022. ISBN 9780906321782
- Allen, P., D. Van Dusen, J. Lundy, and S. Gliessman. 1991. Integrating social, environmental, and economic issues in sustainable agriculture, American Journal of Alternative Agriculture, 6(1):34-19.
- Barker, Joel. 1993. Paradigms, The Business of Discovering the Future, HarperCollins Publisher, New York, NY.
- Canadian Almanac & Directory. Copp Clark Publishing Company. 1847. ISBN 978-1-895021-81-3.
- Curriculum|TOKYO UNIVERSITY OF AGRICULTURE. www.nodai.ac.jp. Retrieved 2021-05-02.
- Davis, John H.; Goldberg, Ray A. (1957). A Concept of Agribusiness. Division of Research, Graduate School of Business Administration, Harvard University. ISBN 9781684225248.
- Definition of irrigation | Dictionary.com. www.dictionary.com. Retrieved 2021-01-21.
- DeLind, Laura. 1994. Organic farming and the social context: A challenge for us all, American Journal of Alternative Agriculture, 9(4):146-147.

- Department of Agribusiness Management and Entrepreneurship. College of Economics and Management. Retrieved 2021-05-02.
- Dixie, Grahame (2007). Horticultural Marketing, Marketing Extension Guide 5 (PDF). FAO, Rome.
- Dobbs, T. and J. Cole. 1992. Potential effects on rural economies of conversion to sustainable farming systems, American Journal of Alternative Agriculture, 7(1&2):70-80.
- Dobbs, T.L., M.G. Leddy, and J.D. Smolik. 1988. Factors influencing the economic potential for alternative farming systems: case analysis of South Dakota, American Journal of Alternative Agriculture 3(1):26-32.
- Drilon, Jose D. (1971). Agribusiness Management Resource Materials: Introduction to agribusiness management. Asian Productivity Organization. ISBN 978-92-833-1009-9.
- Drucker, Peter. 1989. Post Capitalist Society. Harper Business, HarperCollins, Publishers, Inc. New York, New York.
- Drucker, Peter. 1989. The New Realities. Harper and Row Publishers, Inc. New York, New York.
- Dunham, Denis. 1994. Food cost review, 1993, Economic Research Service.
- Guide to the Collections: Henry G. Gilbert Nursery and Seed Trade Catalog Collection. National Agricultural Library, Special Collections. Retrieved April 23, 2009.
- Hamilton, Shane (2016). Revisiting the History of Agribusiness. Business History Review (2016 ed.). 90 (3): 541–545. doi:10.1017/S000768051600074X. ISSN 0007-6805. S2CID 157756414.
- HBS Professor Ray Goldberg on the History of Sustainable Agribusiness, archived from the original on 2021-12-12, retrieved 2021-05-02
- Helen Markelova and Ruth Meinzen-Dick Collective action and market access for smallholders: A summary of findings (PDF). Archived from the original (PDF) on 2010-07-16. Retrieved 2009-01-15. CAPRi/IFPRI 2007

INDEX

www.ingramcontent.com/pod-product-compliance
Lightning Source LLC
Chambersburg PA
CBHW071330230926
51849CB00002B/360
9789349964198